Plant Biotechnology

Springer
Berlin
Heidelberg
New York
Barcelona
Hong Kong
London
Milan
Paris
Singapore
Tokyo

**J. Hammond, P. McGarvey,
and V. Yusibov (Eds.)**

Plant Biotechnology

New Products and Applications

With 12 Figures and 9 Tables

Springer

JOHN HAMMOND, Ph.D.
USDA-ARS, USNA
Florist and Nursery Crops Unit
Rm. 238, B-010A
10300 Baltimore Ave.
Beltsville, MD 20705
USA

PETER MCGARVEY, Ph.D.
National Biomedical Research Foundation
Georgetown University Medical Center
LR-3, Preclinical Science Building
3900 Reservoir Road, NW
Washington, DC 20007
USA

Professor VIDALDI YUSIBOV, Ph.D.
Biotechnology Foundation Labs.
Thomas Jefferson University
Rm. M85, JAH, 1020 Locust Street
Philadelphia, PA 19107
USA

Cover Illustration: The expression of pharmaceuticals and vaccines in transgenic plants and from genetically engineered plant viruses (molecular farming) are symbolized by a plant-containing syringe, virus particles, and antibody molecules.

Cover Design: design & production GmbH, Heidelberg

ISBN 3-540-66265-0 Springer-Verlag Berlin Heidelberg New York

Second Printing 1999 (originally published in CURRENT TOPICS IN MICROBIOLOGY AND IMMUNOLOGY, Vol. 240, ISBN 3-540-65104-7, 1999)

Cataloging-in-Publication Data applied for

Die Deutsche Bibliothek - CIP-Einheitsaufnahme

Plant Biotechnology : new products and applications ; with 9 tables / ed. by John Hammond ... - 2. printing. - Berlin ; Heidelberg ; New York ; Barcelona ; Hong Kong ; London ; Milan ; Paris ; Singapore ; Tokyo : Springer, 1999
 ISBN 3-540-66265-0

Typesetting: Scientific Publishing Services (P) Ltd, Madras

Production Editor: Angélique Gcouta

SPIN: 10739350   27/3020 – 5 4 3 2 1 0 – Printed on acid-free paper

Preface

The title of this volume, *Plant Biotechnology: New Products and Applications*, may look a little out of place among previous volumes of Current Topics in Microbiology and Immunology that have focused mostly on issues related to human health and animal biology. However, plant biology has always been of immense practical importance, and has enjoyed an intimate relationship with medicine and other biological sciences for centuries. Increasing scientific specialization and the dramatic advances in the medical and chemical sciences during this century have left many persons with the impression that plant biology and plant biotechnology is important only in relation to the agricultural sciences. This is no longer true. Within the past year a genetically engineered plant virus has been used to vaccinate and protect against an animal disease (see the chapter by Lomonossoff and Hamilton), the first human trials of a potential transgenic plant-based oral vaccine against cholera have been conducted (see the chapter by Richter and Kipp), and the first human trial of an injectable transgenic plant-derived therapeutic protein is under way (discussed in the chapter by Russell et al.). Today plant biotechnology is being used in new and creative ways to produce therapeutic products for medicine and plastics for industry as well as new disease- and stress-resistant crops for agriculture.

This volume is intended to introduce some of these new and exciting areas to readers unfamiliar with plant biotechnology and also to serve as a review of ongoing research and future considerations for readers familiar with the field. Since it would be impractical to cover all areas of plant biotechnology in a single volume, we have elected to focus on several of the exciting areas of research in which comparatively little has been published. However, we have included review chapters that touch on most of the current research in plant biotechnology and the basic technology employed to engineer plants and plant viruses.

The first chapter by Hammond surveys the field of plant biotechnology from the initial work on pathogen resistance to the current attempts to produce pharmaceuticals and industrial enzymes in plants. This is an excellent chapter for readers interested in learning about the wide variety of research accomplishments in plant biotechnology and how some of the research discussed in the following chapters fits into the overall status of the field.

The chapter by Hansen and Chilton reviews *Agrobacterium*-mediated transformation, the most commonly used method for engineering dicotyledonous crops. Agrobacterium have evolved a system for infecting wounded plants and inserting genes into the plant nucleus. Some of these genes cause tumor formation and other genes produce opines (amino acid analogues) that the bacteria can then use as an energy source. Microbiologists have adapted this system by deleting the genes that cause tumor formation and opine production and replacing them with recombinant DNA of the researchers choice. Advantages of *Agrobacterium*-mediated transformation include the typical insertion of one or only a few copies of the transgene into actively transcribed regions of the plant genome.

The chapter by Finer et al. covers the second major method of transformation, particle bombardment or biolistic transformation, which is generally favored for transformation of monocotyledonous plant species. Particle bombardment is also used for some dicotyledonous plants that do not regenerate efficiently after *Agrobacterium*-mediated transformation. Advantages of particle bombardment are that it can be adapted to most all plant species. However, particle bombardment often results in complex transgene insertion loci, which may cause gene silencing in some instances.

The chapter by Yusibov et al. discusses a new method of transiently engineering plants for the expression of recombinant proteins using genetically engineered plant viruses. This method does not change the genetic makeup of the host plant. Plant viruses engineered to contain recombinant genes have expressed high levels of medically and industrially important proteins in the field. In addition to the expression of full-length proteins, viral capsid proteins have the potential to express small peptides on the virions surface. The chapter by Yusibov et al. and that by Lomonossoff and Hamilton each describe how the expression of peptide antigens on the surface of plant viruses can be used to vaccinate and protect animals against viral infections.

The chapter by Cramer et al. contains an excellent discussion of the potential benefits and challenges in using transgenic plants for production of therapeutic proteins. Cramer et al. discuss two different types of production systems that were designed for ease of postharvest processing. In the first, transgene expression during the growing season is minimized by having the gene regulated by a strong wound-inducible promoter. Thus any potentially deleterious effect on the plant of transgene expression is avoided, no energy is wasted by the crop in producing a product that might be degraded prior to harvest, and greater biomass can be attained. The fresh leaf tissue is harvested to meet demand or is stored under refrigeration until needed. Transgene expression is induced a few hours prior to extraction and processing, so that all of the purified product is recently synthesized. The second approach discussed by Cramer et al. is the fusion of the product of interest to a seed-storage protein of *Brassica napus*; in this system the fusion protein is targeted to the oil bodies that accumulate in rapeseed following expression from a seed-specific promoter. Rapeseed (and the oil bodies therein) are stable for years, with minimal storage requirements.

Russell examines the various factors that must be considered in the expression of antibodies in transgenic plants, from expression levels to glycosylation, and from bioequivalency to regulatory considerations. Indeed, the first human clinical trial of an injectable transgenic plant-derived therapeutic protein is currently under way. Russell also factors in some cost analyses that must be considered, for example, time-to-market figures, and compares the regulatory burden for therapeutic products to that for transgenic plants designed for food use. Russell also addresses the agronomic and plant-breeding strategies that can be applied once a promising transgenic line has been selected, in order to both stabilize the gene and maximize production. Finally, Russell discusses compliance with good manufacturing practices, which is the basis for preparing an application for a product to be accepted for pharmaceutical use.

The chapter by Richter and Kipp provides an excellent review of research using transgenic plants directly as oral vaccines without additional purification steps. Indeed, phase I clinical trials have been completed in which human subjects ate transgenic potatoes containing a potential oral vaccine against cholera. In addition to using transgenic plants to vaccinate against infectious pathogens, there is evidence that they can be used to induce immune tolerance that may prove effective against autoimmune diseases.

The chapter by Tumer et al. discusses a slightly more traditional relationship between medicine and plant biology in which plants provide useful drugs for medicine. Pokeweed antiviral protein belongs to a class of proteins called ribosome-inactivating proteins that are used by plants as a defense against pathogens. Many of these proteins have demonstrated medical applications against animal pathogens. Plant biotechnology allows us to rapidly investigate the function of a potential medicine and to produce it in high quantities in crop plants.

The advances of the past decade may herald a new "Agricultural Revolution" that will rival the "Green Revolution" of the twentieth century. Further developments in plant biotechnology will no doubt follow those described here. The ability to control gene expression in specific tissues, and especially to induce expression in response to externally applied stimuli, are likely to become increasingly precise. The number of species that can be routinely transformed will certainly grow, and can be expected to include all of the major agronomic crops. This will allow increased choice of production areas to fill specific markets and greater selection of storage organs for the expression of products with distinct processing requirements. A further benefit of improved spatial and temporal control of gene expression could be the production of two or more commodities from different tissues of the same crop. The ability to manipulate multiple genes will allow transfer of increasingly complex pathways from one species to another, and allow utilization of biochemistry *in planta* to supplant some of the complexities and pollution from chemical plants. Plants that resist disease or predation, without the application of pesticides that also affect beneficial insects, are another example of how agricultural biotechnology has the potential to bring about an environmentally friendly "Agricultural Revolution".

Peter McGarvey

List of Contents

List of Contributors

(Their addresses can be found at the beginning of their respective chapters.)

Overview: The Many Uses and Applications of Transgenic Plants

J. Hammond

1 Introduction

In this volume we provide background information on the principles, practices, and common methods used for generation of transgenic plants, together with selected examples of the potential for transgenic plants and engineered plant viruses to produce high-value products. The fields of transgenic plants and engineered viruses are expanding so rapidly that it is not possible to cover all of the areas being examined. We have therefore selected chapters addressing only a few areas in detail. Some other areas have been addressed recently, and we review briefly some of these topics in this introductory chapter.

United States Department of Agriculture, Agricultural Research Service, United States National Arboretum, Floral and Nursery Plants Research Unit, Beltsville, MD 20705-2350, USA

2 Plant Transformation

Stable expression of introduced genes over the life of the plant and heritable expression in the progeny require integration of the gene and all of the regulatory sequences into the plant genome. In general it is desirable for the transgene to be integrated into the nuclear DNA, although the plastid genome may be an appropriate target for some constructs. There are many variations of plant transformation methods to integrate the desired gene(s) into the plant genome; these methods form the backbone of plant biotechnology. The most widely used methods are *Agrobacterium*-mediated gene transfer, and transformation mediated by particle bombardment. Advances in methodology are such that it is now possible to transform many crop species that were initially considered recalcitrant. Recent improvements will likely make possible the transformation of all major crops on a routine basis.

The chapter by Hansen and Chilton reviews *Agrobacterium*-mediated transformation – the most commonly used method for most dicotyledonous crops. Different *Agrobacterium* species and isolates may be more effective with different plant species, although *Agrobacterium rhizogenes* may cause many undesirable phenotypic effects in transformed plants (e.g., TEPFER 1984; MAUREL et al. 1991). Advantages of *Agrobacterium*-mediated transformation include the typical insertion of one or a few copies of the transgene, often into actively transcribed regions of the plant genome. Another advantage is the size of the piece of DNA that can be transferred. To date up to 150kb of foreign DNA has been introduced as a single fragment, using binary bacterial artificial chromosome vectors (HAMILTON et al. 1996). More commonly 1–10kb is introduced from a binary plasmid system. Combining the features of *Agrobacterium*-mediated transformation with wounding by particle bombardment may increase the efficiency of infection and/or the recovery of transformed plantlets (BIDNEY et al. 1992). However, *Agrobacterium* does not infect most monocotyledonous species efficiently, and thus other methods are preferred.

The chapter by Finer et al. covers the other major method of transformation, particle bombardment or biolistic transformation, which is generally favored for transformation of most monocotyledonous species. Particle bombardment is also preferred for some dicotyledonous species that do not regenerate efficiently after *Agrobacterium*-mediated transformation. They also compare different methods of particle bombardment to various other transformation methods that have been applied by multiple laboratories. Particle bombardment often results in complex transgene insertion loci, which causes gene silencing in some instances. Finer and colleagues have shown that multiple plasmids can be cotransformed into a single target tissue (HADI et al. 1996), and thus there is the possibility of introducing similarly complex amounts of DNA as with binary bacterial chromosome vectors by *Agrobacterium* transformation. However, with increasing numbers of plasmids to be integrated, there is increasing likelihood of undesirable effects on regulation of the transgenes. Thus the method of transformation that is selected usually depends upon ability to introduce the DNA of interest to a suitable tissue and to regenerate stably transformed plants from the transformed tissues.

A few reports detail methods that do not require tissue culture for recovery of transgenic plants, thus allowing production of transgenics from species for which efficient and reliable tissue culture systems are not available. The chapter by FINER et al. reviews the microtargeting device of SAUTTER et al. (1991), in which a mixture of DNA and particles is used to bombard meristematic tissue in shoot apices. In theory the meristem itself could be transformed, and nonchimeric shoots could then be excised and rooted; alternatively, seed could be collected from flowers arising from transformed tissue. Using similar logic, GRIESBACH (1994) has shown that transgenic plants can be produced by electrophoresing DNA into meristematic tissue of orchid protocorms. At least transient electrophoretic transformation has been reported for axillary meristems of chrysanthemum, carnation, and lisianthus (BURCHI et al. 1995), and stable transformation of peppers (*Capsicum*) demonstrated (R.J. Griesbach, personal communication). Even woody species such as the redbud (*Cercis canadensis*) and plum (*Prunus domestica*) yield shoots in which the introduced DNA can be detected weeks after treatment; however, the technique is labor intensive, and many shoots prove to be tissue chimeras that are not stable (R.J. Griesbach, personal communication).

The chapter by Russell addresses the agronomic and plant breeding strategies that can be applied once an initial transgenic line with a promising expression level has been selected, in order to both stabilize the gene and maximize production. This includes a discussion of isolation distances in different crops – a consideration common among plant breeders, but not necessarily among genetic engineers.

3 Plant Protection

One of the earliest fields in which transgenic plants were exploited can be broadly described as plant protection against biotic stresses, i.e., pests and pathogens. The concept of pathogen-derived resistance was described by SANFORD and JOHNSTON (1985), building on observations of the interactions between bacteria and bacteriophage, and the premise that inappropriate expression of a pathogen gene would interfere with the establishment of infection or the disease process.

Pathogen-derived resistance was rapidly applied to plant viruses, with considerable success, although analysis of the results has shown that there are many different mechanisms that can account for resistance. Some of these resistance mechanisms appear to be related to the ability of plants to regulate aberrant gene expression (SMITH et al. 1994; ENGLISH et al. 1996). Viral genes that have been shown to confer effective resistance to particular viruses include coat protein, replicase, and defective movement proteins (BEACHY et al. 1990; COOPER et al. 1995; LOMONOSSOFF 1995; BAULCOMBE 1996; HAMMOND 1996). Antisense RNA and untranslatable RNA have also been shown to be highly effective against some viruses (LINDBO and DOUGHERTY 1992a,b; HAMMOND and KAMO 1995; TABLER et al. 1998). Nonviral genes have also been shown to confer resistance against several viral or viral-like pathogens. Double-stranded RNA, which serves as either the genome or replicative form of RNA viruses and is not found in healthy eukaryotic cells, can thus be targetted for nonspecific virus resistance (WATANABE et al. 1995; MITRA et al. 1996; HAMMOND 1997). The expression of pokeweed antiviral protein or other ribosome-inactivating proteins may prove to be the most effective broad-spectrum approach for resistance to plant viral and fungal pathogens yet proposed (Tumer et al., this volume)

Resistance to fungal (e.g., CORNELISSEN and MELCHERS 1993; DIXON et al. 1996) and bacterial pathogens (e.g., CARMONA et al. 1993), has typically been achieved by interfering with the integrity of the pathogen cell wall or cell membranes. A number of plant- or bacterial-derived enzymes have been overexpressed in plants and shown to confer varying degrees of protection against infection by different fungi. These include: chitinases and glucanases (e.g., ZHU et al. 1994); seed expressed antimicrobial proteins (e.g., LEAH et al. 1991) that also protect against some bacterial infections (e.g., CARMONA et al. 1993); lysozyme (DÜRING et al. 1992); and silkmoth cecropin (e.g., FLORACK et al. 1995). Other lytic proteins and peptides (e.g., BROKAERT et al. 1995; JAYNES et al. 1993) have also been expressed in transgenic plants and tested against bacterial diseases (DÜRING et al. 1992; HUANG et al. 1997; JAYNES et al. 1993).

Insect resistance produced by expression of insecticidal proteins has become a widely used strategy. One advantage of transgenic expression over spray application is that only those insects that feed on the crop are targeted; another advantage is reduction in the number of equipment passages through the crop. Beneficial insects such as pollinators and predatory species are not affected by the plant-expressed protein – thus maintaining populations of desirable organisms as well as reducing the environmental burden of pesticides. Problems with expression levels in first-generation transgenic plants expressing the *Bacillus thuringensis* CryIA protein led to mutation of the bacterial gene sequence in order to remove cryptic splice sites (e.g., KOZIEL et al. 1993, 1996), and to alter codon usage to more closely reflect that typical of the plant species (MURRAY et al. 1989; PERLAK et al. 1991; KOZIEL et al. 1996). Plant-derived insecticidal proteins examined include proteinase inhibitors such as the cowpea trypsin inhibitor (which conferred resistance to two species of stem borer that attack rice; XU et al. 1996), other proteinase inhibitors, and various lectins (reviewed by BOULTER 1993).

Herbicide resistance genes (e.g., DE BLOCK et al. 1987) can be used as selectable markers for identification of transformed plants (in place of antibiotic resistance). Introduction of a herbicide resistance gene may also be a goal in itself (DE BLOCK et al. 1987), as superior weed control can be obtained even when the most troublesome weeds have similar physiology to the nontransgenic crop. There may again be some environmental advantage, as this can allow the use of herbicides that are rapidly inactivated in the soil, in contrast to other chemicals that persist longer in the environment. A related application of herbicide resistance is the control of parasitic weeds such as broomrape (*Orobanche* spp.) and witchweed (*Striga* spp.), which are major problems in many developing countries (GRESSEL et al. 1996).

4 Plant Quality

Various types of transgenes have been shown to affect crop quality, including a variety of genes that regulate ripening. The most notable case is that of tomato, into which a number of different genes have been introduced that delay postharvest over-ripening, and thus allow fruit to remain on the plant longer and ripen more naturally. Vine ripening of tomato significantly improves flavor compared to fruit picked when immature and ripened artificially after shipping. Several types of transgenic tomato have been produced commercially (see KRIDL and SHEWMAKER 1996; KRAMER and REDENBAUGH 1994; GRIERSON and FRAY 1994). Another approach to improved flavor is the expression of the sweet protein monellin in both tomato and lettuce (PEÑARRUBIA et al. 1992). Starch biosynthesis has been successfully manipulated in tomato (for paste production) and potato (for processing quality and reduced oil uptake) by expression of a bacterial ADP glucose pyrophosphorylase that is insensitive to feedback regulation (reviewed by STARK et al. 1996).

Qualitative changes in oil composition have been reported in several crops. Chain length and the degree of saturation of fatty acids in the oil determines the properties and use of the oil (KRIDL and SHEWMAKER 1996). For example, longer chain fatty acids with a lower degree of polyunsaturates are desirable for frying oils that do not oxidize as quickly, whereas oils high in monounsaturates are better for salad oils; other combinations are desired for margarine or confectionary uses, and novel oils or combinations may have particular industrial applications (KRIDL and SHEWMAKER 1996; METZ and LASSNER 1996). Most of the oil engineering has been carried out in canola, which itself was derived from high erucic acid rapeseed by classical breeding methods. Genetic engineering has further increased the range of "designer" oils that can be obtained from the crop. Rapeseed is already a major oil crop in many parts of the world, with an established oil-processing industry. The ease of transformation, availability of seed-specific promoters, and prior availability of germplasm that has oils with distinct fatty acid profiles makes rapeseed a good target for precise fatty acid modifications to yield oils with new properties (METZ and LASSNER 1996).

Another area that has received attention is improvement in seed protein quality. In many parts of the world grains constitute the major protein source. Grains are typically low in one or more essential amino acids, and efforts have therefore been made to increase their content to improve nutritional value. One approach is to modify naturally occurring seed storage proteins (see HABBEN and LARKINS 1995). Another is to introduce genes for storage proteins rich in essential amino acids from other species (e.g., SUN et al. 1996). This is an area that must be approached with some caution. A methionine-rich protein from Brazil nut was chosen to boost the nutritional value of soybean seed protein, and was shown to be appropriately expressed and inherited (SAALBACH et al. 1994). However, some persons are allergic to Brazil nut, and the protein transferred to soybean is largely responsible for this adverse reaction (NORDLEE et al. 1996). It is thus unlikely that transgenic seed expressing the Brazil nut protein will be commercialized. A thorough examination of the properties of candidate proteins is therefore recommended in order to avoid potential allergenicity or toxicity problems; similar consideration should be given to other proteins to be expressed in edible plants (LEHRER et al. 1996, 1997).

5 New Products

The availability of seed-specific promoters has formed the basis for a whole new category of crops engineered to express desirable products in seed. Seeds are the ultimate storage organ of the plant and are evolved to withstand desiccation. Thus proteins expressed in seed may be stored, following normal agricultural practices, and then purified when desired. This approach is discussed in detail by CRAMER et al. (this volume). Corn (e.g., HOOD et al. 1997; and the chapter by RUSSELL), tobacco (e.g., GANZ et al. 1996) and canola (see chapter by CRAMER et al.) are the main crops examined to date. HOOD et al. (1997) report commercial production of avidin from transgenic maize – perhaps the first case in which a plant-produced heterologous protein has been purified from field-grown plants and offered for sale. They cite several advantages of the transgenic seed source over the natural source in chicken egg whites. These include: the cost of raw materials; reduced volume of material to be processed; stability in the stored seed; and reasonable stability during processing (HOOD et al. 1997). As in other cases, the avidin gene has been modified to use those codons more common in other maize genes. Seed nutritive quality can be improved by increasing the content of essential amino acids expressed in storage proteins (e.g., SUN et al. 1996) or by increasing bioavailability of phosphorus through the expression of phytase in transgenic seed (PEN et al. 1993; VERWOERD and PEN 1996; BEUDEKER 1996).

Phytase-expressing seed may have beneficial effects beyond increased efficiency of phosphorus utilization. Chicken manure rich in phosphorus has been blamed for farm runoff causing eutrophication of waterways in the eastern United States, and possibly for outbreaks of *Pfiesteria piscicida*, which have caused significant fishkills and economic losses. The phosphorus content of manure could be reduced by phytase added to seed. Another approach is using phytase-expressing canola seed as an additive in place of phytase produced from fermentation cultures of *Aspergillus niger*, the current commercial source. The product is expected to be sold as milled seed, thus preventing any germinable seed from reaching the customer. By this means the company retains control over the germplasm and also avoids some regulatory control over genetically modified organisms, as only the processed product and not viable seed is transported across state or national borders (BEUDEKER 1996). Alternately, phytase could be directly expressed in the primary feed grain rather than in an additive. The farmer or feed supplier would not have to mix the additive with nontransgenic grain, which would be an advantage for transgenic feed over added phytase from transgenic canola. However, as feed grains are frequently supplied whole, there would be the potential for loss of control over the germplasm unless the gene were supplied in F1 hybrid varieties. In the case of transgenic F1 hybrid varieties, plants grown from farmer-saved seed would be less productive and more variable, discouraging illegal use of the germplasm.

Other possibilities for expression in transgenic plants are polymers for plastics production, as a renewable alternative to petroleum- or coal-based production. An additional benefit of biopolymers is the potential biodegradation of the resultant plastics, and reduced environmental pollution. Polyhydroxyalkanoates are produced by various bacteria as a storage compound but are quite expensive to produce by fermentation. It is now possible to transfer the necessary pathways to transgenic plants (POIRIER et al. 1992). Initial tests showed that cytoplasmic expression had deleterious effects on plants, but targeting the product to the chloroplast resulted in yields of up to 4% of dry weight as polyhydroxybutyrate with relatively minor effects on plant growth (POIRIER et al. 1992).

6 Pharmaceuticals

Perhaps the greatest economic potential for value-added traits from transgenic plants will come from the expression of biopharmaceuticals. Among the products that are likely to be produced in transgenic plants are cytokines, hormones, monoclonal antibodies (or their derivatives), enzymes, and vaccines (MIELE 1997). Some of these products may be expressed either from stably transformed plants (see chapters by Cramer et al., Richter and Kipp, Russell, and Tumer et al.), or from transient expression systems in the form of recombinant plant viral vectors (described in chapters by Lomonossoff and Hamilton, and Yusibov et al.). Plant viral vectors may also function as test vehicles for the expression of constructs designed for stable transformation (to ensure stability or functionality of engineered proteins, or targeting to the correct subcellular compartment) before going through the longer process of producing and selecting transgenic plants.

Three chapters in this volume relate largely to the use of plant systems for the production of vaccines. As noted in the chapters by Richter and Kipp, Lomonossoff and Hamilton, and Yusibov et al., plant-produced vaccines have several potential advantages. Plants have been used as a source for many pharmaceuticals, and thus there are established purification protocols for the removal of various potentially toxic contaminants from plant extracts; furthermore, plants are not host to many pathogens that can occur in mammalian or avian cell culture systems currently used for vaccine production. Vaccine production usually requires extensive infrastructure for production and administration, including cold-chain maintenance from production to final immunization. The high cost of vaccine production and administration limit the applicability of vaccines in many parts of the world. Vaccine materials that could be grown and prepared locally, in the form of edible plant tissues or viral display vectors, have the potential to make vaccination against many debilitating diseases more accessible to much of the world's population. Initial human trials of plant-expressed vaccines are under way to determine efficacy.

One question that must be addressed before embarking on a program to express a pharmaceutical protein in plants is whether the potential advantages of plant expression are sufficient to make it economically viable over present production systems. There are situations in which the present costs of production are prohibitive. For example, GANZ et al. (1996) point out that recombinant human granulocyte-macrophage colony-stimulating factor (GM-CSF) has been found to be clinically beneficial in several applications, including bone marrow transplantation, cancer chemotherapy, treatment of AIDS- or drug-induced neutropenia, and after major surgery. Other potential applications have been identified but remain to be tested. The current production of GM-CSF from mammalian cell culture is too costly, and the availability too limited, to allow evaluation in additional applications (GANZ et al. 1996). Similar restrictions apply to many other proteins currently produced from tissues such as human placentae or from mammalian recombinant systems. Production in transgenic plants allows dramatic scale up of production, hopefully with equally dramatic reductions in cost. CRAMER et al. (1996) estimated that leaf tissue from a single tobacco plant could yield sufficient recombinant human glucocerebrosidase for a standard dose that would require 500–2000 placentae – and with a significant reduction in biomass to be processed.

Another question concerns the continuity of production and the ability to produce enough to meet increased demand. Many crops can be produced year round by utilizing different geographic locations. Costs and avenues for transportation of agricultural products are relatively inexpensive and well established. Agricultural production can be rapidly increased by planting greater acreage. Proteins produced in seed should be particularly stable and easily stored, allowing flexibility in timing of processing (PEN 1996; and the chapter by Cramer et al.). The leaf tissue postharvest expression system described by Cramer et al. is also well-adapted to processing flexibility. However, there may be variation in both quantity and quality of yield from crops in different locations and seasons, and even between different tissues of the same crop. PEN (1996) discusses both quantitative and qualitative differences in glycosylation that may result from variation in environmental conditions affecting enzyme expression; VERWOERD and PEN (1996) have noted that the degree of glycosylation of phytase in leaf tissue of transgenic tobacco differs from that in seed from the same plants.

Differences in the glycosylation patterns in plants is a major issue in producing biopharmaceuticals (discussed by MIELE 1997; and in chapters by Russell, and by Cramer et al. in this volume). Tissue- and species-specific variation in glycosylation may cause some difficulties, especially if they lead to changes in antigenicity. Most mammalian systems add sialic acid side chains that may increase protein half-life within the mammalian bloodstream. CHRISPEELS and FAYE (1996) have described the problem in detail, and they discuss possibilities for producing plants with altered glycosylation patterns as a result of engineering the enzymatic pathways.

Retention of recombinant proteins in the endoplasmic reticulum (ER) avoids the protein being modified in the golgi apparatus. Some proteins are retained in the ER as a consequence of the C-terminal retention signals KDEL or HDEL (PELHAM 1990), but addition of these signals does not ensure ER retention. A second possibility is to select or create plants with mutations in the golgi enzymes involved in modification of N-linked glycans. Mutants of *Arabidopsis* lacking an essential enzyme in the modification pathway have been isolated by VON SCHAEWEN et al. (1993); these plants do not incorporate any complex glycans but are phenotypically normal. It should thus be possible to engineer other modifications in the glycan pathways, including the addition of sialic acid side chains (CHRISPEELS and FAYE 1996).

Results reported in the chapter by Russell, and by others, suggest that plant-produced proteins typically retain activity comparable to proteins produced in yeast or eukaryotic cell culture systems, and that altered antigenicity is not a significant problem. Indeed, plant-produced vaccine candidates are recognized by antibodies induced by wild-type proteins (HAQ et al. 1995; MCGARVEY et al. 1995). DALSGAARD et al. (1997) have demonstrated that a plant-produced vaccine confers effective protection on animals immunized with plant virus particles bearing the engineered immunogenic sequence. GANZ et al. (1996) observed that plant-produced human GM-CSF is apparently not glycosylated, despite carrying potential sites for both N- and O-linked glycosylation. The unmodified plant-produced protein was shown to have equal biological activity to GM-CSF produced in mammalian cell culture (GANZ et al. 1996).

Characteristics other than glycosylation may be affected by the production system. OLINS and LEE (1993) reviewed expression of heterologous proteins in *Escherichia coli* and noted that both mistranslation and posttranslational modifications result in significant levels of amino acid heterogeneity in recombinant proteins. The degree of mistranslation is correlated with expression levels, and may be affected by anomalous codon usage (PEN 1996). Foreign proteins expressed in *E. coli* frequently retain the initiating (N-terminal) methionine residue, which may affect the biological activity, and require either a redesign to cause secretion and cleavage or in vitro enzymatic cleavage (PEN 1996).

Tumer et al. (this volume) discuss the function and potential applications of naturally occurring plant-expressed antiviral proteins, concentrating on pokeweed antiviral protein. This is perhaps the chapter with the closest relationship to the traditional preparation of pharmaceuticals from plants; they present the history and current state of understanding of the antiviral activity of ribosome-inactivating proteins (RIPs). They describe the use of RIPs to create virus-resistant transgenic plants, and separation of the antiviral activity from plant toxicity. RIPs have also been shown to have activity against mammalian viruses, including HIV. As discussed by Tumer et al., other RIPs such as trichosanthin have also been expressed in transgenic plants (LAM et al. 1996); trichosanthin has potential utility in treatment of HIV and other human viral diseases and has also been expressed from plant viral vectors (KUMAGAI et al. 1993; see also chapter by Yusibov et al.).

7 Safety and Quality Control Issues

Production of diagnostic or therapeutic products in plants has many potential advantages. Although there is at least one broad-range pathogen (*Pseudomonas aeruginosa*) that can cause disease in both plants and animals under experimental conditions (RAHME et al. 1995), and some toxin-producing bacteria and fungi may occur on plants, plants are not hosts to major mammalian pathogens that can contaminate cell-culture systems for production. Plants can synthesize many compounds that are beyond the ability of present techniques of organic chemistry to synthesize economically. Many pharmaceuticals were first identified from plants used in traditional medicine. Thus many pharmaceuticals are presently, or were originally, prepared from plant substrates (MIELE 1997), and there is thus considerable experience in purification of pharmaceuticals from plant tissues on an industrial scale. Skill in removal of potential contaminants such as alkaloids or other plant metabolites already exists. Considerations related to genetic stability, microbiological contamination, purity, and comparability to products from other production systems have been reviewed by MIELE (1997).

The chapter by Russell examines factors to be considered in expression of antibodies in transgenic plants. These include: expression levels; glycosylation; bio-equivalency; and regulatory considerations. Differences in the glycosylation pattern may affect both antibody half-life and antigenicity; the significance of such differences at the practical level is not yet clear, although the first human clinical trial of an injectable transgenic plant-derived therapeutic protein is currently underway and may provide some answers. Russell also compares the regulatory burden for therapeutic products to that for transgenic plants designed for food use, and discusses compliance with good manufacturing practices – the basis for preparing an application for a product to be accepted for pharmaceutical use.

Input costs for growing plants are lower than for many other production methods such as fermentation (HOOD et al. 1997), and scale-up of production is very rapid. PEN (1996) discusses various factors affecting choice of production of recombinant proteins in transgenic plants in comparison to bacteria, yeast, filamentous fungi, insect or mammalian cells in culture, or transgenic animals. These factors include: required production levels, development times, downstream processing, and product characteristics. Downstream processing of proteins from transgenic plants has been discussed in detail by HATTI-KAUL and MATTIASSON (1996) and compared to microbial cell and animal tissue production systems. This is a valuable adjunct to the discussions of the specific instances and considerations described in the chapters by Cramer et al. and Russell. Comparative costs of production in plants are discussed by HIATT (1990), PEN (1996), PARMENTER et al. (1996), and HOOD et al. (1997).

8 Transgenic Plants vs. Viral Vectors

It is not always necessary, nor even desirable, to produce stably transformed plants. The chapters by Yusibov et al. and Lomonossoff and Hamilton address expression of proteins from plant viral vectors. Viral replication amplifies many-fold any gene that is engineered into a suitable position in the plant viral genome. Plant viruses can allow much more rapid expression of significant quantities of proteins in inoculated plants and allow testing of multiple constructs before proceeding with stable transformation. Viruses can spread rapidly within inoculated plants, and viral coat proteins often make up a significant proportion of the total protein in infected cells. Introduced genes may be expressed as translational fusions with a viral protein or from viral subgenomic promoters, depending upon the required properties. Replacement of a portion of the viral coat protein gene may be a highly valuable means of producing vaccines. The modified coat protein assembled into virions presents the epitope as a regular array that can stimulate the immune system more effectively than a subunit vaccine (see chapter by Lomonossoff and Hamilton). Translational fusions to viral coat proteins could be purified as modified virions, and the added domain(s) cleaved with appropriate proteases to allow separation from the viral protein (as discussed for oleosin fusions in the chapter by Cramer et al.).

One difference between viral vectors and stably transformed plants is the location in which the desired product is accumulated. In most cases proteins expressed from plant viruses accumulates primarily in the leaf tissue and must be harvested from fresh tissue or tissue stored frozen. In contrast, products can be produced in specific tissues of transgenic plants. Depending upon the promoters used, the product can be produced in seed or tubers that can be stored for extended periods without sophisticated equipment and still retain a high degree of integrity and activity. Products produced from both viral vectors or transgenic plants can be targeted to different subcellular locations, allowing for posttranslational modifications or avoiding aggregation problems, and the product is subject to the same kind of modifications (e.g., proteolytic processing and glycosylation) if the appropriate signals are present (see below). Scale-up of virally expressed products can be much faster, provided that suitable plants are available for infection   and production can be switched between products simply by virtue of inoculating plants with one modified virus in place of another.

The earliest plant viral vectors were based on cauliflower mosaic virus (CaMV), a double-stranded DNA virus that replicates through an RNA intermediate (GRONENBORN et al. 1981; BRISSON et al. 1984). One of the earliest applications for expression from a plant virus of a protein with pharmaceutical potential was interferon expression from CaMV, although the aim in this case was to determine the effect of interferon on an RNA plant virus *in planta* (DE ZOETEN et al. 1989). The insertion capacity of CaMV has since been shown to be quite limited, and RNA virus vectors have become more commonly used, despite early doubts as to their usefulness and stability (VAN VLOTEN-DOTING et al. 1985).

Recombinant genes translated from subgenomic RNAs have been used to express a variety of pharmaceutical proteins (see chapter by Yusibov et al.). These proteins have been produced at high yields, and Biosource Technologies is working on protocols for processing commercial quantities of leaf tissue to obtain drug grade product.

Products in stably transformed transgenic plants are translated from RNA transcribed from the host genomic DNA. Virally expressed genes are typically transcribed into infectious RNA from modified cDNA clones and after initial infection are replicated through a dsRNA intermediate. As RNA viruses lack a proof-reading mechanism, there is thus greater opportunity for introduction of mutations into the replicating virus. In most cases, however, the inserted sequence has proven no more mutable than the viral sequence, and has shown stability over multiple host passages and considerable time periods (KEARNEY et al. 1993). The stability of early viral vectors was a problem, and the insert was frequently lost from the virus after several host passages or more than a few weeks in a single plant. The integration of additional genes often somewhat reduces the fitness of the engineered virus. Since fitness is often recovered when the introduced gene is deleted, over time the proportion of the engineered virus may be reduced to undetectable levels. This may be useful for environmental containment of engineered viruses (DELLA-CIOPPA and GRILL 1996). Second-generation vectors have improved the stability significantly (see chapter by Yusibov et al.), although some constructs may still be noticeably less stable than others. This may be a significant quality control issue.

9 Localization and Processing of Expressed Proteins

Depending upon the protein to be expressed and its stability or requirement for processing, plant-expressed proteins can be targeted to different subcellular or extracellular locations. Proteins that require glysosylation need to be targeted to the ER, and then possibly to the golgi apparatus for further modification (see discussion of glycosylation above). If it is desired to retain the protein in the ER, it is not necessary for plant signal sequences to be utilized, as it has been shown that at least some bacterial, fungal, and mammalian ER signal sequences are functional in plants (ASPEGREN et al. 1996; MA and HIATT 1996; and references therein). CONCEIÇÃO and RAIKHEL (1996) have reviewed the accumulation of soluble proteins via the secretory system, the signals that determine retention in the ER versus transport to the vacuole, and proteolytic processing in the vacuole.

Cramer et al. discuss two different types of production that were designed for ease of postharvest processing. In the first, transgene expression during the crop growing season is minimized by having the gene regulated by a strong wound-inducible promoter active primarily in leaf tissue. Thus any potentially deleterious effect on the plant of transgene expression is avoided; no energy is wasted by the crop in producing product that might be degraded prior to harvest, and greater biomass can be attained. The fresh leaf tissue is harvested either to meet demand, or harvested and stored under refrigeration until needed. Transgene expression is then induced a few hours prior to extraction and processing, so that all of the purified product is recently synthesized. There is less chance of product degradation from such induced tissues than in tissues in which product is synthesized continually – and separation of production of the recombinant product from environmental influences on crop growth. Other approaches along similar lines include organ-specific expression from promoters active only in appropriate growth stages, such as tuber-specific expression in potato or ethylene-induced expression in ripening tomato fruit.

Control of expression can also be obtained during the growing season by the use of inducible promoters responding to an externally supplied stimulus. Such a system would ideally have a low basal activity and a high level of activity following induction, reverting to basal levels after the stimulus is removed. Several systems have been described that are not applicable at the field level, or that are insufficiently repressed for adequate control of gene expression. These include tetracycline-dependent expression (e.g., GATZ 1996), or copper-induced expression (METT et al. 1993) which may lead to phytotoxicity problems. Perhaps the most promising system for field application reported to date is the ethanol-inducible promoter from *Aspergillus nidulans*; this system appears to be tightly regulated and environmentally acceptable (CADDICK et al. 1998).

The second major approach discussed by Cramer et al. is the fusion of the product of interest to a seed-storage protein of *Brassica napus*. In this system the fusion protein is targeted to the oil bodies that accumulate in rapeseed following expression from a seed-specific promoter. Rapeseed (and the oil bodies therein) are stable for years with minimal storage requirements; the oil bodies remain quite stable following aqueous extraction, after which they can be separated from soluble contaminants by flotation centrifugation. The recombinant product can then be partitioned into the aqueous phase following site-specific enzymatic cleavage from the seed-storage protein, which remains embedded in the oil body.

Each technique for expression has advantages for particular applications. The postharvest expression using the wound-inducible promoter might be chosen when posttranslational processing such as glycosylation is required, or if endomembrane targeting is necessary. The oil body separation process utilizes existing industrial equipment that can be readily adapted for cost-effective production of recombinant proteins. Indeed, oil body expression systems have already been identified as having potential for production of immobilized protein matrices, which have applications in many processes (KÜHNEL et al. 1996).

MA and HIATT (1996) have described the assembly of multi-subunit antibodies following passage through the ER in plants and subsequent secretion through the cell membrane into the apoplastic space, in which hydrolytic processing is minimal. By transforming individual plants with separate immunoglobulin chains and sequentially crossing these plants, they were able to generate offspring expressing all combinations from individual chains, through monomeric, dimeric, and secretory immunoglobulin molecules (MA and HIATT 1996). It should therefore be possible to correctly assemble other multi-subunit products in transgenic plants prior to purification.

In contrast, if intracellular expression of antibody is required, it is more effective to express antibody fragments that do not require passage through the ER for correct assembly to attain a functional conformation. Intracellular expression is more likely to result in biologically functional antibody fragments *in planta*, as demonstrated by OWEN et al. (1992). OWEN et al. expressed a functional single chain antibody against phytochrome A, which alters light-mediated responses affecting germination. Another example is the demonstration by TAVLADORAKI et al. (1993) of reduced virus infection and a delay in symptom development in plants expressing an antiviral single chain antibody. Seed expression may be more appropriate for antibody and antibody fragments purified for pharmaceutical use or industrial applications, as discussed in the chapter by Russell.

10 Bioremediation

A further application of plant produced antibodies is in bioremediation (OWEN et al. 1996), whereby either antibodies or antibody fragments would sequester the pollutants on the surface of plant cells (in the apoplastic space), or catalytic antibodies would mediate the breakdown of the pollutant to nontoxic products. In the former scenario, the plant tissue in which the pollutant is concentrated would be harvested for treatment or recovery under more controlled conditions ex situ, while the latter approach would permit on-going in situ remediation. The existence of nontransgenic plants capable of growing in heavily polluted soils such as mine tailings has been exploited to stabilize spoil tips, and even for metal reclamation from such sites (SMITH and BRADSHAW 1979). Transgenic plants with introduced capacity to bind heavy metal ions are a newer development (PAN et al. 1994). Choosing an appropriate species of transgenic plants may allow more efficient harvesting and processing of biomass to facilitate recovery of commercial levels of metals from materials that are currently hazardous wastes.

11 A Note of Caution

An issue that must be faced by all who work with transgenic plants in other than a purely academic setting is the question of intellectual property rights. There are multiple levels at which such considerations apply to transgenic plants. The first question that should be considered is the ownership of the germplasm into which the trait of interest is to be introduced. In most cases it is desirable to select a variety of a crop plant that has already been selected on the basis of its agronomic properties. These might include: the ability to perform well in particular geographic locations, responsiveness to fertilizers, resistance to pests and diseases, and suitability for storage and processing. There are some varieties that are in the public domain, but many varieties with desirable characteristics are covered by plant variety protection certificates or plant patents (KJELDGAARD and MARSH 1996), and agreement of the breeder or seed company (in the form of licencing) will be required for commercialization of a genetically engineered derivative.

Further rights are associated with almost every aspect of the production of transgenic plants. There are multiple patents relating to methods of transformation, to particular vectors, the selectable markers, and to particular genes or classes of gene constructs. A review of the issues involved was published by HUSKISSON (1996), but a list of relevant patents and patent holders is far too extensive for meaningful discussion. Any such list would also be very quickly out of date as a result of newly issued patents and court decisions regarding interferences. There are also the cases of so-called "submarine" patents, which may have been initially filed several years ago, but of which the scientific community remains ignorant until the issued patent surfaces.

There is no legal requirement that a patentee must licence a covered invention (either item or process) to any party who requests a licence to use or practice the invention. Whereas it is common practice for holders of related patents to cross-licence their inventions in order to avoid expensive legal battles over patent infringement, it is also quite common for patentees to refuse licences or delay negotiations for a licence in order to gain or maintain a competitive advantage. Requested licence fees may also be beyond the capability of smaller companies; conversely a smaller company in possession of a patent may not have the resources to prosecute a larger concern that infringes upon that patent.

It should therefore come as no surprise that there has been a considerable reorganization of the plant biotechnology industry, with many smaller biotechnology companies being taken over by the larger companies. Many seed companies have become part of much larger groups, as companies seek to own a broader base of germplasm in which to express their proprietary genes and thus control the market all the way from germplasm to value-added product. There will likely continue to be a place for start-up companies to introduce new ideas and products, but many such businesses will either be forced into strategic alliances with the industry giants or will be acquired by one.

References

Aspegren K, Mannonen L, Ritala A, Teeri TH (1996) Production of fungal, heat-stable β-glucanase in suspension cultures of transgenic barley cells. In: Owen MRL, Pen J (eds) Transgenic plants: a production system for industrial and pharmaceutical proteins. Wiley, Chichester, pp 201–212

Baulcombe DC (1996) Mechanisms of pathogen-derived resistance to viruses in transgenic plants. Plant Cell 8:1833–1844

Beachy RN, Loesch-Fries S, Tumer NE (1990) Coat protein-mediated resistance against virus infection. Annu Rev Phytopathol 28:451–474

Beudeker RF (1996) Commercialization of phytase-containing seeds. In: Owen MRL, Pen J (eds) Transgenic plants: a production system for industrial and pharmaceutical proteins. Wiley, Chichester, pp 329–337

Bidney D, Scelonge C, Martich J, Burrus M, Sims L, Huffman G (1992) Microprojectile bombardment of plant tissues increases transformation frequency by Agrobacterium tumefaciens. Plant Mol Biol 18:301–313

Boulter D (1993) Insect pest control by copying nature using genetically engineered crops. Phytochemistry 34:1453–1466

Brisson N, Paszkowski J, Penswick JR, Gronenborn B, Potrykus I, Hohn T (1984) Expression of a bacterial gene in plants by using a viral vector. Nature 310:511–514

Broekaert WF, Terras FRG, Cammue BPA, Osborn RW (1995) Plant defensins: Novel antimicrobial peptides as components of the host defence system. Plant Physiol 108:1353–1358

Burchi G, Griesbach RJ, Mercuri A, De Benedetti L, Priore D, Schiva T (1995) In vivo electrotransfection: transient GUS expression in ornamentals. J Genet Breeding 49:163–168

Caddick MX, Greenland AJ, Jepson I, Krause, K-P, Qu N, Riddell KV, Salter MG, Schuch W, Sonnewald U, Thomsett AB (1998) An ethanol inducible gene switch for plants used to manipulate carbon metabolism. Nature Biotechnology 16:177–180

Carmona MJ, Molina A, Fernández JA, López-Fando JJ, García-Olmedo F (1993) Expression of the α-thionin gene from barley in tobacco confers enhanced resistance to bacterial pathogens. Plant J 3: 457–462

Chrispeels MJ, Faye L (1996) The production of recombinant glycoproteins with defined non-immunogenic glycans. In: Owen MRL, Pen J (eds) Transgenic plants: a production system for industrial and pharmaceutical proteins. Wiley, Chichester, pp 99–113

Conceição ADa S, Raikhel NV (1996) Accumulation of soluble proteins in the endomembrane system of plants. In: Owen MRL, Pen J (eds) Transgenic plants: a production system for industrial and pharmaceutical proteins. Wiley, Chichester, pp 75–98

Cooper B, Lapidot M, Heick JA, Dodds JA, Beachy RN (1995) A defective movement protein of TMV in transgenic plants confers resistance to multiple viruses whereas the functional analog increases susceptibility. Virology 206:307–313

Cornelissen, BJC, Melchers LS (1993) Strategies for control of fungal diseases with transgenic plants. Plant Physiol 101:709–712

Cramer CL, Weissenborn DL, Oishi KK, Radin DN (1996) High-level production of enzymatically active human lysosomal proteins in transgenic plants. In: Owen MRL, Pen J (eds) Transgenic plants: a production system for industrial and pharmaceutical proteins. Wiley, Chichester, pp 299–310

Dalsgaard K, Uttenthal Å, Jones TD, Xu F, Merryweather A, Hamilton WDO, Langeveld JPM, Boshuizen RS, Kamstrup S, Lomonossoff GP, Porta C, Vela C, Casal JI, Meloen RH, Rodgers PB (1997) Plant-derived vaccine protects target animals against a viral disease. Nat Biotechnol 15:248–252

De Block M, Botterman J, Vanderwiele M, Dockx J, Thoen C, Gosselé V, Rao Movva N, Thompson C, Van Montagu M, Leemans J (1987) Engineering herbicide resistance in plants by expression of a detoxifying enzyme. EMBO J 6:2513–2518

De Zoeten GA, Penswick JR, Horisberger MA, Ahl P, Schultze M, Hohn T (1989) The expression, localization, and effect of a human interferon in plants. Virology 172:213–222

Della-Cioppa G, Grill LK (1996) Production of novel compounds in higher plants by transfection with RNA viral vectors. Ann NY Acad Sci 792:57–61

Dixon RA, Lamb CJ, Paiva NL, Masoud S (1996) Improvement of natural defense responses. Ann NY Acad Sci 792:126–139

Düring K, Fladung M, Lörz H (1992) Antibacterial resistance of transgenic potato plants producing T4 lysozyme. In: Nester EE, Verma DPS (eds) Advances in molecular genetics of plant-microbe interactions. Kluwer, Dordrecht, pp 573–577

English JJ, Mueller E, Baulcombe DC (1996) Suppression of virus accumulation in transgenic plants exhibiting silencing of nuclear genes. Plant Cell 8:179–188

Florack D, Allefs S, Bollen R, Bosch D, Visser B, Stiekema W (1995) Expression of giant silkmoth cecropin B encoding genes in transgenic tobacco. Transgenic Res 4:132–141

Ganz PR, Dudani AK, Tackaberry ES, Sardana R, Sauder C, Cheng X, Altosaar I (1996) Expression of human blood proteins in transgenic plants: the cytokine GM-CSF as a model protein. In: Owen

MRL, Pen J (eds) Transgenic plants: a production system for industrial and pharmaceutical proteins. Wiley, Chichester, pp 281–297

Gatz C (1996) Chemically inducible promoters in transgenic plants. Curr Opin Biotechnol 7:168–172

Gressel J, Ransom JK, Hassan EA (1996) Biotech-derived herbicide-resistant crops for Third World needs. Ann NY Acad Sci 792:140–153

Grierson D, Fray R (1994) Control of ripening in transgenic tomatoes. Euphytica 79:251 263

Griesbach RJ (1994) An improved method for transforming plants through electrophoresis. Plant Sci 102:81–89

Gronenborn B, Gardner RC, Schaefer S, Shepherd RJ (1981) Propagation of foreign DNA in plants using cauliflower mosaic virus as vector. Nature 294:773–776

Habben JE, Larkins BA (1995) Genetic modification of seed proteins. Curr Opin Biotechnol 6:171 174

Hadi MZ, McMullen MD, Finer JJ (1996) Transformation of 12 different plasmids into soybean via particle bombardment. Plant Cell Rep 15:500–505

Hamilton CM, Frary A, Lewis C, Tanksley SD (1996) Stable transfer of intact high molecular weight DNA into plant chromosomes. Proc Natl Acad Sci USA 93:9975–9979

Hammond J (1996) Biotechnology and resistance. Acta Horticulturae 432:246 256

Hammond J (1997) Repelling plant pathogens with ribonuclease. Nat Biotech 15:1247

Hammond J, Kamo KK (1995) Effective resistance to potyvirus infection in transgenic plants expressing antisense RNA. Mol Plant-Microbe Interact 8:674–682

Haq TA, Mason HS, Clements JD, Arntzen CJ (1995) Oral immunization with a recombinant bacterial antigen produced in transgenic plants. Science 268:714–716

Hatti-Kaul R, Mattiasson B (1996) Downstream processing of proteins from transgenic plants. In: Owen MRL, Pen J (eds) Transgenic plants: a production system for industrial and pharmaceutical proteins. Wiley, Chichester, pp 115–147

Hiatt A (1990) Antibodies produced in plants. Nature 344:469–470

Hood EE, Witcher DR, Maddock S, Meyer T, Baszczynski C, Bailey M, Flynn P, Register J, Marshall L, Bond D, Kulisek E, Kusnadi A, Evangelista R, Nikolov Z, Wooge C, Mehigh RJ, Hernan R, Kappel W, Ritland D, Li CP, Howard JA (1997) Commercial production of avidin from transgenic maize: characterization of transformant, production, processing, extraction and purification. Mol Breeding 3:291–306

Huang Y, Nordeen RO, Di M, Owens LD, McBeath JH (1997) Expression of an engineered cecropin gene cassette in transgenic tobacco plants confers disease resistance to *Pseudomonas syringae* pv. *tabaci*. Phytopathology 87:494–499

Huskisson FM (1996) Patents and biotechnology. In: Owen MRL, Pen J (eds) Transgenic plants: a production system for industrial and pharmaceutical proteins. Wiley, Chichester, pp 313 328

Jaynes JM, Nagpala P, Destéfano-Beltrán L, Huang JH, Kim J, Denny T, Cetiner S (1993) Expression of a Cecropin B lytic peptide analog in transgenic tobacco confers enhanced resistance to bacterial wilt caused by *Pseudomonas solanacearum*. Plant Sci 89:43–53

Kearney CM, Donson J, Jones GE, Dawson WO (1993) Low level of genetic drift in foreign sequences replicating in an RNA virus in plants. Virology 192:11–17

Kjeldgaard RH, Marsh DR (1996) Recent United States developments in plant patents. Mol Breeding 2:95–96

Koziel MG, Beland GL, Bowman C, Carozzi NB, Crenshaw R, Crossland L, Dawson J, Desai N, Hill M, Kadwell K, Launis K, Lewis K, Maddox D, McPherson K, Meghji MR, Merlin E, Rhodes R, Warren GW, Wright M, Evola SV (1993) Field performance of elite transgenic maize plants expressing an insecticidal protein derived from *Bacillus thuringiensis*. Biotechnology 11:194 200

Koziel MG, Carozzi NB, Desai N, Warren GW, Dawson J, Dunder E, Launis K, Evola SV (1996) Transgenic maize for the control of European corn borer and other maize insect pests. Ann NY Acad Sci 792:164–171

Kramer MG, Redenbaugh K (1994) Commercialization of a tomato with an antisense polygalacturonase gene: the FLAVRSAVR tomato story. Euphytica 79:293–297

Kridl JC, Shewmaker CK (1996) Food for thought: Improvement of food quality and composition through genetic engineering. Ann NY Acad Sci 792:1 12

Kühnel B, Holbrook LA, Moloney MM, van Rooijen GJH (1996) Oil bodies of transgenic *Brassica napus* as a source of immobilized β-glucuronidase. J Am Oil Chem Soc 73:1533 1538

Kumagai MH, Turpen TH, Weinzettl N, Della-Cioppa G, Turpen AM, Donson J, Hilf ME, Grantham GL, Dawson WO, Chow TP, Piatak M Jr, Grill LK (1993) Rapid high-level expression of biologically active α-trichosanthin in transfected plants by an RNA viral vector. Proc Natl Acad Sci USA 90: 427–430

Lam YH, Wong B, Wang RN-S, Yeung HW, Shaw PC (1996) Use of trichosanthin to reduce infection by turnip mosaic virus. Plant Sci 114:111–117

Leah R, Tommerup H, Svendsen I, Mundy J (1991) Biochemical and molecular characterization of three barley seed proteins with antifungal properties. J Biol Chem 266:1564–1573

Lehrer SB, Horner WE, Reese G (1996) Why are some proteins allergenic? Implications for biotechnology. Crit Rev Food Sci Nutr 36:553–564

Lehrer SB, Reese G (1997) Recombinant proteins in newly developed foods: identification of allergenic activity. Int Arch Allergy Immunol 113:122–124

Lindbo JA, Dougherty WG (1992a) Pathogen-derived resistance to a potyvirus: Immune and resistant phenotypes in transgenic tobacco expressing altered forms of a potyvirus coat protein nucleotide sequence. Mol Plant-Microbe Interact 5:144–153

Lindbo JA, Dougherty WG (1992b) Untranslatable transcripts of the tobacco etch virus coat protein gene sequence can interfere with tobacco etch virus replication in transgenic plants and protoplasts. Virology 189:725–733

Lomonossoff GP (1995) Pathogen-derived resistance to plant viruses. Annu Rev Phytopathol 33:323–343

Ma, J K-C, Hiatt A (1996) Expressing antibodies in plants for immunotherapy. In: Owen MRL, Pen J (eds) Transgenic plants: a production system for industrial and pharmaceutical proteins. Wiley, Chichester, pp 229–243

Maurel C, Barbierbrygoo H, Spena A, Tempé J, Guern J (1991) Single rol genes from the *Agrobacterium rhizogenes* TL-DNA alter some of the cellular responses to auxin in *Nicotiana tabacum*. Plant Physiol 97:212–216

McGarvey PB, Hammond J, Dienelt MM, Hooper DC, Fu ZF, Dietzschold B, Koprowski H, Michaels FH (1995) Expression of the rabies virus glycoprotein in transgenic tomatoes. Biotechnology 13:1484–1487

Mett V, Lochhead LP, and Reynolds PHS (1993) Copper-controllable gene expression system for whole plants. Proc Natl Acad Sci USA 90:4567–4571

Metz J, Lassner M (1996) Reprogramming of oil synthesis in rapeseed: Industrial applications. Ann NY Acad Sci 792:82–90

Miele L (1997) Plants as bioreactors for biopharmaceuticals: regulatory considerations. Trends Biotechnol 15:45–50

Mitra A, Higgins DW, Langenberg WG, Nie H, Sengupta DN, Silverman RH (1996) A mammalian 2–5A system functions as an antiviral pathway in transgenic plants. Proc Natl Acad Sci USA 93:6780–6785

Murray E, Lotzer J, Eberle M (1989) Codon usage in plants. Nucleic Acids Res 17:477–498

Nordlee JA, Taylor SL, Townsend JA, Thomas LA, Bush RK (1996) Identification of a Brazil-nut allergen in transgenic soybeans. N Engl J Med 334:688–692

Olins PO, Lee SC (1993) Recent advances in heterologous gene expression in *Escherichia coli*. Curr Opin Biotechnol 4:520–525

Owen MRL Gandecha A, Cockburn W, Whitelam GC (1992) Synthesis of a functional anti-phytochrome single-chain Fv protein in transgenic tobacco. Biotechnology 10:790–794

Owen MRL, Cockburn W, Whitelam GC (1996) The expression of recombinant antibody fragments in plants. In: Owen MRL, Pen J (eds) Transgenic plants: a production system for industrial and pharmaceutical proteins. Wiley, Chichester, pp 245–260

Pan A, Yang M, Tie F, Li L, Chen Z, Ru B (1994) Expression of mouse metallothionin-I gene confers cadmium resistance in transgenic tobacco plants. Plant Mol Biol 24:341–351

Parmenter DL, Boothe JG, Moloney MM (1996) Production and purification of recombinant hirudin from plant seeds. In: Owen MRL, Pen J (eds) Transgenic plants: a production system for industrial and pharmaceutical proteins. Wiley, Chichester, pp 261–280

Pelham HRB (1990) The retention signal for soluble proteins of the endoplasmic reticulum. Trends Biochem Sci 15:483–486

Pen J (1996) Comparison of host systems for the production of recombinant proteins. In: Owen MRL, Pen J (eds) Transgenic plants: a production system for industrial and pharmaceutical proteins. Wiley, Chichester, pp 149–168

Pen J, Verwoerd TC, van Paridon PA, Buedeker RF, van den Elzen PJM, Geerse K, van der Klis JD, Versteegh JAJ, van Ooyen AJJ, Hoekema A (1993) Phytase-containing transgenic seed as a novel feed additive for improved phosphorus utilization. Biotechnology 11:811–814

Peñarrubia L, Kim R, Giovannoni J, Kim, S-H, Fischer RL (1992) Production of the sweet protein monellin in transgenic plants. Biotechnology 10:561–564

Perlak FJ, Fuchs RL, Dean DA, McPherson SL, Fischhoff DA (1991) Modification of the coding sequence enhances plant expression of insect control protein genes. Proc Natl Acad Sci USA 88:3324–3328

Poirier YP, Dennis DE, Klomparens K, Somerville CR (1992) Production of polyhydroxybutyrate, a biodegradable thermoplastic, in higher plants. Science 256:520–523

Rahme LG, Stevens EJ, Wolfort SF, Shao J, Tompkins RG, Ausubel FM (1995) Common virulence factors for bacterial virulence in plants and animals. Science 268:1899–1902

Saalbach I, Pickardt T, Machemehl F, Saalbach G, Schneider O, Müntz, K (1994) A chimeric gene encoding the methionine-rich 2 s albumin of the Brazil nut (*Bertholletia excelsa* H.B.K.) is stably expressed and inherited in transgenic grain legumes. Mol Gen Genet 242:226–236

Sanford JC, Johnston SA (1985) The concept of parasite-derived resistance deriving resistance genes from the parasite's own genome. J Theor Biol 113:395–405

Sautter C, Waldner H, Neuhaus-Url G, Galli A, Niehaus G, Potrykus I (1991) Microtargeting: high efficiency gene transfer using a novel approach for the acceleration of micro-particles. Biotechnology 9:1080–1085

Smith RAH, Bradshaw AD (1979) The use of metal tolerant plant populations for the reclamation of metalliferous waste. J Appl Ecol 16:595–612

Smith HA, Swaney SL, Parks TD, Wernsman EA, Dougherty WG (1994) Transgenic plant virus resistance mediated by untranslatable sense RNAs: Expression, regulation, and fate of non-essential RNAs. Plant Cell 6:1441–1453

Stark DM, Barry GF, Kishore GM (1996) Improvement of food quality traits through enhancement of starch biosynthesis. Ann NY Acad Sci 792:26–36

Sun SSM, Zuo W, Tu HM, Xiong L (1996) Plant proteins: engineering for improved quality. Ann NY Acad Sci 792:37–42

Tabler M, Tsagris M, Hammond J (1998) Antisense RNA- and ribozyme-mediated resistance to plant viruses. In: Hadidi A, Khetarpal RK, Koganezawa E (eds) Plant viral disease control. APS, St. Paul. pp 79–93

Tavladoraki P, Benvenuto E, Trinca S, De Martinis D, Cattaneo A, Galeffi P (1993) Transgenic plants expressing a functional single-chain Fv antibody are specifically protected against virus attack. Nature 366:469–472

Tepfer D (1984) Transformation of several species of higher plants by *Agrobacterium rhizogenes*: sexual transmission of the transformed genotype and phenotype. Cell 47:959–967

van Vloten-Doting L, Bol J F, Cornelissen BJM (1985) Plant-virus-based vectors for gene transfer will be of limited use because of the high error frequency during viral RNA synthesis. Plant Mol Biol 4:323–326

Verwoerd TC, Pen J (1996) Phytase produced in transgenic plants for use as a novel feed additive. In: Owen MRL, Pen J (eds) Transgenic plants: a production system for industrial and pharmaceutical proteins. Wiley, Chichester, pp 213–225

Von Schaewen A, Sturm A, O'Neill J, Chrispeels MJ (1993) Isolation of a mutant arabidopsis plant that lacks N-acetyl glucosaminyl transferase I and is unable to synthesise Golgi-modified complex N-linked glycans. Plant Physiol 102:1109–1118

Watanabe Y, Ogawa T, Takahashi H, Ishida I, Takeuchi Y, Yamamoto M, Okada Y (1995) Resistance against multiple plant viruses in plants mediated by a double stranded-RNA specific ribonuclease. FEBS Lett 372:165–168

Xu D, Xue Q, McElroy D, Mawal Y, Hilder VA, Wu R (1996) Constitutive expression of a cowpea trypsin inhibitor gene, CpTi, in transgenic rice plants confers resistance to two major rice insect pests. Mol Breeding 2:167–173

Zhu Q, Maher EA, Masoud S, Dixon RA, Lamb CJ (1994) Enhanced protection against fungal attack by constitutive co-expression of chitinase and glucanase genes in transgenic tobacco. Biotechnology 12:807–812

Lessons in Gene Transfer to Plants by a Gifted Microbe

G. HANSEN and M.D. CHILTON

Novartis Agribusiness Biotechnology Research, Inc., 3054 Cornwallis Road, Research Triangle Park, NC 27709, USA

1 Foreword

The discovery that the bacterial phytopathogen *Agrobacterium* is a natural expert at interkingdom gene transfer has proven to be rewarding in areas far afield from plant pathology. It has provided researchers with a powerful means of dissecting and eventually modifying the genomes of many kinds of plants. It has led to landmark discoveries about the molecular basis of plant-pathogen interactions. It has furthered our understanding of plant gene function and regulation involved in the physiological and developmental processes in plants. *Agrobacterium* has played a key role in many of the recent advances in plant science, and future exploitation of this microbe is limited only by the imagination of its scientific practitioners. Access to a wide array of genes with good potential for crop improvement is now intensifying interest in development of a plant transformation system with consistent, optimal and cost-effective performance. Once again, *Agrobacterium* seems destined for a central role.

The study of this micro-organism has formed the basis of such a variety of research and technical applications that a truly comprehensive review would be of encyclopedic size. In this chapter we present an update on the most recent advances in the molecular studies of *Agrobacterium*. This review focuses on recent efforts to improve *Agrobacterium* T-DNA transfer as a transformation system for "recalcitrant crops", with emphasis on information useful to new practitioners in this arena. For additional information we refer the reader to earlier reviews with different emphasis (OTTEN et al. 1992; HOOYKAAS and SCHILPEROORT 1992; VAN DE BROEK and VANDERLEYDEN 1995; ZUPAN and ZAMBRYSKI 1995; SHENG and CITOVSKY 1996; ZAMBRYSKI 1992; KONCZ et al. 1994; ROSSI et al. 1998). We have also supplemented these with additional suggestions at relevant points in the text.

2 Introduction

A. tumefaciens and *A. rhizogenes*, soil micro-organisms that inhabit the plant rhizosphere, incite crown gall tumors and hairy root disease, respectively, on a wide range of dicotyledonous plants and gymnosperms as well as on some monocotyledonous plants. In nature, the infection process begins at a wound site (LIPPINCOTT and LIPPINCOTT 1969) and leads to formation of tumorous overgrowths. The name crown gall derives from the site of natural galls at the junction of the stem and root (crown), but galls can form at any wound site, including graft unions. In addition to annuals, the disease afflicts numerous woody species such as grapevine, poplar, stone-fruit trees, and ornamental plants (DE CLEENE and DE LEY 1976). Whereas crown gall tissues and hairy roots can be maintained indefinitely in axenic cultures on hormone-free media, uninfected tissues generally need exogenous phytohormones to grow. A large tumor-inducing (Ti) or root-inducing (Ri) plasmid in virulent strains of *Agrobacterium* is accountable for the transformed phenotype of the plant cells (ZAENEN et al. 1974; VAN LAREBEKE et al. 1974; WATSON et al. 1975; WHITE and NESTER 1980). During the process of infection at wound sites in susceptible plants a specific segment of the plasmid DNA called T-DNA (transferred DNA) is mobilized from the bacterium to the plant, where it is covalently joined to plant DNA (CHILTON et al. 1977, 1982). T-DNA is transferred by a conjugation-like process from the bacterium to the host plant cells where it is inserted into random sites in the nuclear genome, apparently favoring genes over heterochromatic regions (see below). Ti plasmid T-DNA contains, among others, genes for synthesis of cytokinin and auxin, phytohormones that cause proliferation of plant cells to form the gall (reviewed in KLEE et al. 1987; REAM 1989). Ri plasmid T-DNAs have diverse oncogenes and are characterized by *rol* (root locus) genes involved in root morphogenesis (WHITE et al. 1985). Plant cells transformed by either Ti or Ri plasmids are usually directed by T-DNA genes to produce novel metabolites called opines, such as octopine and nopaline, that serve as specific nutrients for the inciting bacteria. The characteristic opine produced is often used to type Ti and Ri plasmids (for review see DESSAUX et al. 1991).

The recent introduction of transgenic plants has changed the horizons of agriculture. For such real world application of genetically modified crops, it is essential that transgenes be inherited and expressed uniformly during the many cycles of breeding that precede the release of seeds to growers. In order to facilitate such stability of transgene behavior it would be highly desirable for the transgenic seeds to contain only a single copy of the gene of interest without unnecessary vector or marker genes. Direct DNA delivery methods, including delivery by the Biolistic device, tend to introduce multiple copies of DNA often rearranged into complex arrays and usually inserted at a single site in the plant genome (see Finer et al., this volume). In contrast, *Agrobacterium* often delivers T-DNA as a single copy with boundaries defined by signal DNA sequences on the Ti plasmid. Although the host range of *Agrobacterium* for gall formation initially appeared to be limited mainly to dicots, recent advances have adapted the T-DNA vector system for efficient application in recalcitrant monocots such as maize, rice, and wheat. A further refinement is the cotransformation approach, the delivery of two separate T-DNAs from the same *Agrobacterium* cell. The two T-DNAs are often inserted into unlinked sites in the host plant genome, allowing later genetic segregation. This approach allows the gene(s) of interest and the selectable marker gene to be positioned on separate T-DNAs, affording use of the selectable marker in the initial transformant generation and allowing its elimination during subsequent generations of plant breeding.

The *Agrobacterium* T-DNA delivery system, because of its precision and simplicity, is technically the most attractive method for engineering transgenic plants with foreign DNA integrated into their genome. Recent success in applying *Agrobacterium* gene transfer to a widening range of plants suggests that this vector may be universally useful. It is therefore timely to review recent developments in this field in order to extend the success of the many investigators who have advanced our understanding.

3 Overview of the *Agrobacterium* System

3.1 Location of Genes Involved in *Agrobacterium* T-DNA Transfer

Several genetic components of *Agrobacterium* are required for plant cell transformation, and although in nature they are found on the Ti plasmid or the bacterial chromosome, some have been manipulated onto separate replicons with no impairment of function. First is the pair of border sequences that delineate what *Agrobacterium* recognizes as T-DNA. These are ca. 25-basepair imperfectly matched direct repeats called the left border (LB) and right border (RB). These must be situated on whatever DNA vector (plasmid or other replicon) carries the genes to be transferred (transgenes). Second is a series of about 25 linked virulence (*vir*) genes which are normally situated on the Ti plasmid and arranged in seven operons (STACHEL and NESTER 1986). The exact complement of *vir* genes required may vary slightly depending on the host plant and/or the *Agrobacterium* chromosomal background. One or more *vir* genes may be situated on other replicons in *Agrobacterium* for experimental convenience or to increase their copy number, thereby enhancing their expression level. Third is a series of chromosomal virulence genes that determine attachment to the plant cell, synthesis of cellulose fibrils, and additional functions that are not yet understood.

3.2 Attachment

Attachment of *Agrobacterium* to host plant cells is a prerequisite for DNA transfer. While many attachment (*att*) genes have been elucidated, the mechanism of this intriguing process is not fully understood. A 20-kb block of chromosomal *att* genes is involved in attachment. Mutants *attA1* through *attH* can be complemented by adding conditioned medium (plant culture medium in which carrot cells and wild-type *Agrobacterium* have been grown). Four of these genes exhibit homology to components of ABC transport systems (MATTHYSSE et al. 1996), but it is not clear whether this is a system for uptake or export. Mutant attR, which is not complemented by conditioned medium, is defective in production of an acidic polysaccharide (REUHS et al. 1997). Purified acidic polysaccharide from wild-type *Agrobacterium* can compete with wild-type *Agrobacterium* for attachment to carrot cells or *Arabidopsis* root segments (REUHS et al. 1997).

After the initial attachment step, *Agrobacterium* produces a network of cellulose fibrils that bind the bacterium tightly to the plant cell surface and create a mat that entraps other *Agrobacteria* not yet attached (MATTHYSSE 1986). Cellulose production is important for efficiency of transformation but not an absolute requirement: mutants deficient in cellulose synthesis are 10^{-1}–10^{-3} times as efficient at tumor induction. The genes necessary for cellulose synthesis are known (MATTHYSSE et al. 1995a), and models have been proposed for the mechanism (MATTHYSSE et al. 1995b), but clear proof is still lacking.

Chromosomal virulence genes *chv*A, *chv*B, and *psc*A are also involved in the attachment of *Agrobacterium* to plant cells. ChvB is a 235-kDa protein involved in the formation of a cyclic β-1,2-glucan (ZORREGUIETA and UGALDE 1986) and *chv*A determines the synthesis of a transport protein, located in the inner membrane, necessary for the export of the β-1,2 glucan into the periplasm (CANGELOSI et al. 1989). Pleiotropic effects of *chv*B have been noted, including loss of rhicadhesin activity. Rhicadhesin is an adhesive protein which is thought by some to be directly involved in attachment (SMIT et al. 1992). Virulence can be restored to *chv*B mutants by growth in high osmoticum, a procedure which also restores activity of rhicadhesin. Mutation in *psc*A (*exo*C) also has pleiotropic effects among which is loss of production of β-1,2-glucan (THOMASHOW et al. 1987).

Two plant surface components have been proposed for the role of *Agrobacterium* binding site. One is a vitronectinlike molecule that is present on the surface of many plant cells (WAGNER and MATTHYSSE 1992). Anti-vitronectin antibodies can compete with *Agrobacterium* for attachment to carrot suspension cells (WAGNER and MATTHYSSE 1992). The other proposed binding site for *Agrobacterium* is a plant receptor for rhicadhesin, which SWART et al. (1994a) have identified as the bacterial component of the interaction of *Rhizobiaceae* with plants. SWART et al. (1994b) used rhicadhesin binding to identify and purify a 32-kDa glycoprotein from pea root cell walls that they propose to be the *Agrobacterium* receptor. The different assays employed and/or the difference in the plant studied presumably explain why these two groups arrived at different results. It may be important to understand the identity and bacterial or plant specificity of receptor(s) employed by *Agrobacterium* in order to extend the host range of this vector system as widely as possible. Plant varieties can exhibit resistance to *Agrobacterium* infection by virtue of the absence of receptors: two *Arabidopsis* ecotypes that were identified as refractory to tumor induction failed to bind *Agrobacterium* (NAM et al. 1997).

3.3 Induction of Virulence Functions

Virulence functions are transcriptionally regulated by a two-component gene-regulatory system belonging to a large family of bacterial chemosensors that respond to the chemical environment (LEROUX et al. 1987; DAS 1994). Optimal *vir* gene induction occurs at acidic pH and in the presence of phenolic inducers, such as acetosyringone (AS), that are released by wounded plant cells (STACHEL et al. 1985). The *vir* gene regulatory system operates through two monocistronic virulence genes: *vir*A and *vir*G (reviewed in HOOYKAAS and SCHILPEROORT 1992). The constitutively expressed *vir*A gene produces a protein located in the inner membrane that responds to plant wound metabolites. VirA is a membrane-spanning protein with an N-terminal periplasmic "sensor" domain (senses AS and related phenolics), a "linker" domain (responds to pH and interacts with ChvE, a sugar-binding protein) that includes a membrane-spanning portion of the protein, a "kinase" domain and a "receiver" domain (see Fig. 1). Because of its complex structure, VirA can respond to subtle changes in the environment. At suboptimal AS con-

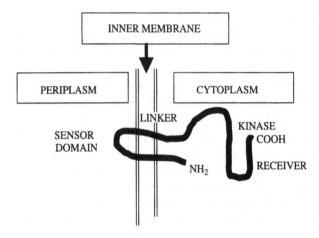

Fig. 1. Schematic of VirA protein

centrations, VirA can be further stimulated by any of a list of sugars, crown gall opines, or amino acids. The resulting autophosphorylation of VirA protein leads to activation of the intracellular VirG, which is phosphorylated at aspartic acid residue 52 by autophosphorylated VirA (JIN et al. 1990) and becomes the transcriptional activator for all *vir* genes including its own. Promoters of *vir* genes possess one or more 12 bp "*vir* box" sequences in their promoter regions (WINANS et al. 1987). A search for mutants that express their *vir* genes constitutively led to the discovery of *virG*-N54D (PAZOUR et al. 1992). This mutation apparently leads to a conformation of the protein that resembles that of phosphorylated wild-type VirG. Constitutive expression of *vir* genes by *Agrobacterium* eliminates the time-consuming process of inducing *Agrobacterium* prior to inoculation onto plant tissue. The *virG*N54D mutant might be more effective in transformation because it is in the induced state at all times. Experiments comparing the efficiency of transformation of a constitutive VirG vector system with that of the corresponding inducible one showed that the mutant is indeed advantageous, especially for recalcitrant plant systems (HANSEN et al. 1994). A caveat about constitutive expression of *vir* genes is the potential for instability of the vector (FORTIN et al. 1993). FORTIN et al. (1992) noted that in some *Agrobacterium* strains, virulence can be lost when the culture is grown in the presence of the *vir* gene inducer acetosyringone. Predominant among mutants identified were IS426 insertions in *virA* or *virG* (FORTIN et al. 1992), which turn off induction. The constitutive mutant *virG* has not been demonstrated to cause similar instability, but it should create similar selection pressure favoring avirulence. Thus one should be alert to such a possibility, especially in strains with IS elements.

A nonessential plant-inducible locus in the *vir* region, called *virH* (or *pinF*), may be involved in the detoxification of some phenolics secreted by plant wounds. Two *virH* products exhibit sequence homology with cytochrome P450 enzymes that catalyze the NADH-dependent oxidation of aromatic substrates (KANEMOTO et al.

1989). This finding suggests that *vir*H proteins may inactivate toxic plant phenolics by a similar mechanism.

3.4 T-DNA Borders

The boundaries of T-DNA are defined by specific ca. 25-bp imperfect direct repeats (YADAV et al. 1982). These are described as 23-, 24-, or 25-bp by various authors depending on the plasmid, but for historical reasons (YADAV et al. 1982) all borders in this chapter are represented on the basis of a 25-bp core structure as defined in the illustration below. Any DNA between these borders is transferred from *Agrobacterium* to the plant cell (reviewed by ZAMBRYSKI 1988). Despite the DNA sequence similarity between left and right borders, studies of border functions have shown that T-DNA borders are differentially utilized. Analysis of the T-DNA content of different transformed plant lines has revealed that the integration of T-DNA into the plant genome takes place at or near these direct repeats (at, on the right; near, on the left) (SLIGHTOM et al. 1986; GHEYSEN et al. 1991; MAYERHOFER et al. 1991; KONCZ et al. 1989). A representation of the bases conserved on the right and left borders of Ti plasmids is shown in Fig. 2, with what we define as the 25-bp core border sequence indicated. Deletion or inversion of the right border sequence results in an almost complete loss of T-DNA transfer (i.e., transfer of the desired DNA left of the right border), whereas deletion of the left border repeat has almost no effect (SHAW 1984; JEN and CHILTON 1986).

Genetic analyses showed that T-DNA transfer is polar and determined by the orientation of the border repeats (ZAMBRYSKI 1992). This is a consequence of the asymmetric cleavage of the T-DNA border sequence (ALBRIGHT et al. 1987): the strand and orientation are switched if the border sequence is inverted, and a different part of the plasmid is then the initial part of T-DNA. The DNA sequence context around the T-DNA borders greatly influences their activity: sequences adjoining the right border enhance, and sequences surrounding the left border reduce, the initiation of polar DNA transfer (WANG et al. 1987a). A *cis*-active sequence of 24 bp, called "overdrive", is present next to the right border of the octopine plasmid (PERALTA et al. 1986). Overdrive stimulates T-DNA transfer even when situated several thousand base pairs away from the border (VAN HAAREN et al. 1987). However, it cannot mediate T-DNA transfer by itself (PERALTA et al. 1986; VAN HAAREN et al. 1987). The overdrive sequence was originally discovered

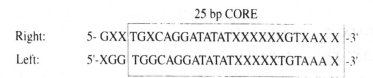

25 bp CORE

Right: 5'- GXX|TGXCAGGATATATXXXXXXGTXAX X|-3'

Left: 5'-XGG|TGGCAGGATATATXXXXXTGTAAA X|-3'

Fig. 2. Conserved bases on the right and left borders of Ti plasmids. *Boxed domain*, what we define as the 25-bp core border sequence indicated. (Adapted from JOUANIN et al. 1989)

in a region to the right of the octopine Ti plasmid TL-DNA right border repeat. Similar sequences are present adjacent to the right border repeat of the octopine pTi TR region and of the agropine pRi TL and TR regions (SLIGHTOM et al. 1986; JOUANIN et al. 1989). Comparison between sequences in the vicinity of the right border revealed a region of 8 bp called a core sequence (5'TGTTTGTT3') located at different distances to the right of each right border repeat. The mannopine-type pRi8l96 T-DNA right border does not contain any sequence identical to the overdrive sequence but instead exhibits an 8-bp related sequence (5'ATTAGTTC3'), repeated six times (HANSEN et al. 1991). This sequence is indeed functionally equivalent to overdrive (HANSEN et al. 1992). A consensus between the central portion of these overdrive signals would be: PuTT(A/T)GTT. No reiterated sequence resembling overdrive has been found near nopaline Ti plasmid T-DNA borders (WANG et al. 1987b).

3.5 T-DNA Processing

After induction of their *vir* genes *Agrobacterium* cells generate a linear single-stranded DNA that is, or matches, the bottom strand of the T-DNA region, designated the T-strand, with its 5' end at the right T-DNA border and 3' end at the left border (STACHEL et al. 1986). T-strand cleavage requires two polypeptides from the *vir*D operon: VirD1 and VirD2 (YANOFSKY et al. 1986; STACHEL et al. 1986; ALT-MOERBE et al. 1986; STACHEL et al. 1987; DE VOS and ZAMBRYSKI 1989; FILICHKIN and GELVIN 1993; PORTER et al. 1987). VirD2 is a site-specific endonuclease (YANOFSKY et al. 1986) that cleaves the lower strand of the border sequences near the left end. The nick was mapped by ALBRIGHT et al. (1987) and by WANG et al. (1987a) to a position at the 3rd or 4th basepair of the 25-bp repeat of octopine T-DNA borders. Later biochemical studies have confirmed that the site of the nick is between 3rd and 4th basepairs of the 25-bp border repeat, as shown in Fig. 3.

An initial report of type I topoisomerase activity for VirD1-containing extracts (GHAI and DAS 1989) was not borne out by later studies of more highly purified VirD1 protein (SCHEIFFELE et al. 1995). The combination of VirD1 and VirD2 in vitro catalyzed T-border-specific nicking, but only for supercoiled plasmid DNA (SCHEIFFELE et al. 1995). Cleavage of single-stranded oligonucleotide border se-

5'-CGGCAGGATATATTCAATTGTAAAT

　　GCCGTCCTATATAAGTTAACATTTA-5'

a

5'- TGGCAGGATATATACCGTTGTAATT

　　ACCGTCCTATATATGGCAACATTAA-5'

b

Fig. 3a,b. Site of the nick between 3rd and 4th basepairs (▲) of the 25-bp border repeat (*underlined*, T-strand portion). **a** Left border. **b** Right border

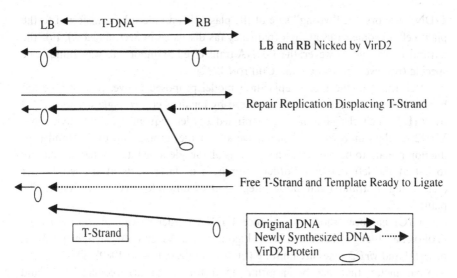

Fig. 4. The possibility of the nick at the right border of T-DNA serving as a starting point for DNA repair synthesis. This would displace the T-strand as it proceeds until the next nick site is reached at its 3' end (left border)

quence required only VirD2 (PANSEGRAU et al. 1993; JASPER et al. 1994). Cutting occurred specifically after the 3rd nucleotide of the 25 bp bottom strand of the left or right border sequence as illustrated above.

VirD2, upon cleaving the border, becomes covalently attached to the 5' end of the T-strand via tyrosine residue 29 (for review, see ZUPAN and ZAMBRYSKI 1995). Since there may be as much as 23 kb of T-DNA between borders of native Ti plasmids (150 kb for BIBACs; HAMILTON et al. 1996), nicking the border sequences does not immediately cause a T-strand to separate from the parent plasmid. It is plausible that the nick at the right border of T-DNA may serve as a starting point for DNA repair synthesis, which would displace the T-strand as it proceeds, until the next nick site is reached at its 3' end (left border) as indicated in Fig. 4.

After repair replication has reached the left border the process can begin again, leading to accumulation of multiple T-strands in a single cell. Although VirD2 protein performs an enzymatic function, it is used in stoichiometric amounts from the point of view of *Agrobacterium*. For this reason, increasing the expression level of *virD2* gene would be expected to improve the efficiency of the vector system, as indeed it does (WANG et al. 1990). Increased *vir* gene expression is a feature of several high efficiency vectors (see section on vectors below).

A problem with the model shown above is that after the posited first VirD1/VirD2-catalyzed nicking reaction at the right border, the T-DNA-containing plasmid should no longer be a substrate for a second nick at the left border because it is no longer supercoiled. Moreover, if the plasmid were nicked at the left border as well as the right, one would expect equal probability for the generation of a "V-strand" (vector strand) consisting of vector DNA extending leftward from the left border, eventually terminating at the right. In the case of artificial "binary"

T-DNA vectors, the "wrong" side of the plasmid can indeed be transferred to the plant cell, in some cases at high frequency (as discussed in Sect. 6 below). For the natural Ti plasmid, however, the T-DNA transferred to tumor cells was found to be specific (reviewed by BEVAN and CHILTON 1982).

Returning to the repair-replication model proposed above, perhaps the left border sequence of the T-strand displaced by repair is cut *as single-stranded DNA* by VirD2. Such cleavage of single-stranded border sequences is an activity that VirD2 exhibits in vitro (see below). Details of the termination of T-strand production remain to be clarified. This point is of considerable interest, because failure to cut at the left border ("border skipping") or false starts at the left border (generating V-strands) can cause integration of unwanted vector DNA into the plant.

VirC1 has been shown to enhance T-DNA border nicking in *Agrobacterium* (TORO et al. 1988) and also in a heterologous *E. coli* system only when the products of *virD1* and *virD2* genes are limiting (DE VOS and ZAMBRYSKI 1989). This suggests two parameters that may be important to vector design: (a) Keeping VirD1 and VirD2 relatively low may be important for correct border identification; (b) VirC1 may associate with VirD1 and VirD2 or with the border repeat and/or at the overdrive sequence to promote right border nicking and T-strand production (TORO et al. 1989). These features unique to the right border region presumably lead to its correct identification as the initiation nick site. It may be desirable, in constructing artificial vector plasmids, to include a natural right border rather than merely a 25-bp oligonucleotide with only the right border repeat sequence. Lacking the flanking DNA, such an artificial right border might serve as initiation point rarely, or only half of the time, leading to frequent delivery of the wrong part of the plasmid to plant cells.

3.6 Transfer to the Plant

After years of controversy it is now generally accepted that T-DNA enters the plant cell nucleus in single-stranded form. Physical evidence that T-DNA is single stranded is that its detection by polymerase chain reaction (PCR) is eliminated by S1 nuclease digestion of the template (YUSIBOV et al. 1994). Genetic evidence for single-stranded T-DNA is based on the extrachromosomal recombination characteristics of T-DNA compared to those of double-stranded and single-stranded DNA (TINLAND et al. 1994).

A more recent topic of controversy is whether T-DNA travels to the plant cell alone (BINNS et al. 1995; SUNDBERG et al. 1996) or as a complex with VirE2 single-strand binding protein as initially proposed (GIETL et al. 1987; CITOVSKY et al. 1988). VirE2 is required for tumor induction: a *virE2* mutant is essentially avirulent. [Note that ROSSI et al. (1996) can measure an extremely low level of transforming activity in a *virE2* deletion mutant.] However, the *virE2* mutation can be complemented by providing VirE2 protein in the host plant through expression of a chimeric transgene (CITOVSKY et al. 1992). The essential function of VirE2 therefore takes place in the plant cell; it is not required for T-strand protection in *Agrobacterium* or for transfer to the plant. As an intriguing alternative, a *virE2*-defective *Agrobacterium* can be complemented by coinoculating the plant cells with a second *Agrobacterium* proficient for VirE2 production but lacking T-DNA (OTTEN et al. 1984). The success of this "extracellular complementation" is dependent upon functional *virB4*, *-5*, *-6*, *-8*, *-9*, and *-10* (BINNS et al. 1995) as well as *virB2*, *-3*, and *-7* (FULLNER 1998) and *virB11* (CHRISTIE et al. 1989) in both *Agrobacterium* donor strains. VirE1 protein is required for transfer of VirE2 but not for T-strand transfer to the plant cell (SUNDBERG et al. 1996). A *virE1* mutant *Agrobacterium* is avirulent. This is ascribed to its inability to deliver VirE2 protein, despite the fact that the bacterium appears to be replete with VirE2 and can transfer T-strands perfectly well (SUNDBERG et al. 1996). From the avirulence of the *virE1* mutant it seems clear that *at least in the absence of the VirE1 protein*, VirE2 protein does not travel to the plant aboard T-strands in sufficient quantity to enable T-DNA to transform the plant cell.

Studies of a nonpolar *virB1* mutation have led FULLNER (1998) to conclude that VirE2 protein can cotransfer with T-strands. In studies of extracellular complementation, Fullner found that a *virB1* mutation interferes differentially with the VirE2-contributing strain. While a *virB1* mutation has only a small effect in the T-strand donor (complementation succeeds: small tumors are formed), the same virB1 mutation in the VirE2 donor completely blocks extracellular complementation. In contrast, a *virB1* mutation does not block tumor induction in an otherwise wild-type cell, i.e., one proficient for VirE2 and T-strand production. The separate transfer model, Fullner (1998) argues, would predict that this strain should be avirulent: the *virB1* mutation would block the (supposedly separate) VirE2 transfer to the plant, just as it blocks the activity of the VirE2 donor in extracellular complementation. This prediction is not fulfilled: otherwise wild-type strain *A348ΔvirB1* induces "attenuated" tumors. A resolution of this seeming contradiction can be visualized, however. The very fact that tumors induced by *A348ΔvirB1* are attenuated (i.e., small) argues that the *virB1* mutation has exerted a strong negative influence on the process, presumably because of its complete inhibition of separate transfer of VirE2. This very finding argues that VirE2 is indeed *mainly* transferred separately in wild-type bacteria. Fullner's results show that infrequently transfer of VirE2 is promoted by T-strands, perhaps by cotransfer. The results of extracellular complementation demonstrate clearly that T-strands and VirE2 *can* travel separately to the plant cell with good efficiency, a conclusion which Fullner's data do not contradict.

By Tn5 insertion mutagenesis of nopaline strain A208, a chromosomal virulence gene called *acvB* has been identified that seems to be involved in the process of T-strand transfer. The gene encodes a single-stranded-DNA-binding protein that is localized in the periplasmic space of *Agrobacterium*, and that binds to T-strands, as detected by immunoprecipitation (WIRAWAN et al. 1993; KANG et al. 1994; WIRAWAN and KOJIMA 1996). The octopine Ti plasmid *virJ* gene encodes a protein 50% identical to *acvB* that can complement an *acvB* mutant. VirJ has generally been found dispensable for tumor induction, presumably because of the presence of *acvB* on the chromosome of the common laboratory strains of *Agrobacterium* used for virulence tests.

The T-complex is transferred to host plant cells through a bacterial transmembrane channel encoded by the *vir*B operon. There are strong similarities between the processing and delivery to the plant cell of T-DNA and the inter-bacterial conjugation of plasmids (for review see LESSL and LANKA 1994; ZUPAN and ZAMBRYSKI 1995). The *vir*B operon consists of 11 genes that are required to transfer the constituents of the T complex through the bacterial inner and outer membrane into the cytoplasm of the host plant cell. Some of the VirB proteins have a high degree of sequence similarity to gene products involved in conjugative functions and with some proteins involved in the mechanisms for protein export. Whereas *virB2* through virB11 genes encode factors that are absolutely essential for DNA transfer to plants (for review, see CHRISTIE 1997), *virB1* seems to play an auxiliary role. As discussed above, *virB1* mutants are virulent, but T-DNA transfer efficiencies typically are reduced by one or two orders of magnitude (BERGER and CHRISTIE 1994). VirB1 possesses motifs that are conserved among eukaryotic lysozymes and bacterial and bacteriophage lytic transglycosylases; this suggests a possible role in the process of pilus assembly, and/or to the formation or stabilization of contacts between *Agrobacterium* and plant recipient cells (BARON et al. 1997a). The subunit of the pilus involved in T-DNA transfer (FULLNER et al. 1996) is a processed form of VirB2 protein (LAI and KADO 1998). VirB4 has a nucleotide binding site and VirB11 is both an ATPase and a protein kinase and could provide energy for the translocation of the T-complex (CHRISTIE et al. 1989). VirB5 is localized to the cytoplasm and inner membrane of *Agrobacterium*, while VirB1, VirB4, VirB8, VirB9, VirB10, and VirB11 are found in membrane fractions (THORSTENSON et al. 1993; FINBERG et al. 1995). VirB7 exists as homodimers and heterodimerized with VirB9 through S–S linkage (BARON et al. 1997b). Models of how all of these pieces may be assembled are still in formative stages.

As the T-strand arrives in the plant cell (presumably the cytoplasm) at some point it complexes with VirE2, and it is plausible that the resulting T-complex recruits plant proteins in its subsequent journey to and through the nuclear membrane. For proteins the process of nuclear import is governed by nuclear localization sequences (NLSs) in the amino acid sequence. T-strand import into the nucleus is mediated by NLSs in one or more of its protein components, VirD2 and/or VirE2. VirD2 protein contains two bipartite NLS structures that have been demonstrated to transport reporter proteins into plant nuclei (HERRERA-ESTRELLA et al. 1990; HOWARD et al. 1992; TINLAND et al. 1992; ROSSI et al. 1993), as does VirE2 (CITOVSKY et al. 1992). It appears that each of these may participate, but no single one is essential to tumorigenesis, at least in laboratory tests. Such seemingly redundant functions may be essential for transformation of other host plants in the field. Interestingly, the NLS of VirD2 has been found to interact with a plant NLS-binding protein belonging to the karyopherin α family (BALLAS and CITOVSKY 1997). VirD2 has also been found to interact with plant cyclophilins, but the interaction is neither with NLS nor ω (see below) domains. The exact role of these proteins is not defined, so their involvement with T-DNA transfer is not yet understood (DENG et al. 1998).

Mutation/deletion of NLSs in VirD2 can reduce transient expression of a T-DNA GUS marker, and can have an impact on the efficiency of tumor induction. Because nuclear uptake may not normally be a rate-limiting step in tumor induction, the effect of a given change in an NLS on nuclear import (as monitored by transient expression) may be considerably greater than its effect on tumor induction. Measuring size of tobacco or potato tumors, SHURVINTON et al. (1992) found that deletion of either one or both of the NLSs of VirD2 reduced tumor induction to 60%–80% of wild-type value. ROSSI et al. (1993), measuring transient expression levels, found that changes in the N-terminal NLS had little or no effect, while altering the C-terminal NLS could reduce GUS expression below 1% of the control value. The latter group ascribes lack of function of the N-terminal NLS to its proximity to the site of VirD2 attachment to the T-strand.

In the course of studying the C-terminal NLS, SHURVINTON et al. (1992) discovered a sequence of amino acids at the C-terminal end of VirD2, named the "ω sequence" because of its location, that is very important for tumor induction: replacement of these amino acids reduced tumor induction to 3% of wild type. Deletion or replacement of the ω sequence (BRAVO-ANGEL et al. 1998) did not affect the fidelity with which the right border of T-DNA was integrated into plant DNA, but transformation was greatly reduced as was transient expression. However, the "efficiency" of integration (estimated as the ratio of transformants to transient expression) was found to be similar to that of wild type. NARASIMHULU et al. (1996) have found that a deletion or deletion/substitution mutation of the ω sequence leads to Agrobacteria that exhibit fairly high transient expression of a T-DNA marker (20%–30% of wild type) but only 0%–3% T-DNA integration. Slight differences in the definition of the ω sequence may account for seeming contradictions and it is also plausible that differing small changes in the C-terminal region of VirD2 may have varied effects on protein conformation. It is clear that VirD2 plays an important role in T-strand formation, transfer, nuclear uptake and integration. It should now be no great surprise that any of these roles might be affected by changes in the ω sequence.

One of the two NLSs of VirE2 overlaps the single-stranded DNA binding domain (CITOVSKY et al. 1992). Deletion of both NLSs to test whether either is functional in nuclear uptake is thus not feasible: without the single-stranded binding domain, activity and virulence are inevitably lost. The evidence that essential VirE2 functions take place in the plant cell (for review, see ZUPAN and ZAMBRYSKI 1995) has been discussed above.

3.7 T-DNA Integration into the Plant Genome

Recently each of the two proteins associated with the T-strand in the plant cell, VirD2 and VirE2, has been implicated directly or indirectly in the integration process. VirD2 protein exhibits in vitro a T-DNA border specific cleavage-and-religation activity analogous to that required for T-DNA excision and ligation to plant DNA (PANSEGRAU et al. 1993). Although it is thus tempting to posit a ligation role for VirD2 in T-DNA integration, no such nonspecific ligating activity has been demonstrated for purified VirD2. On the other hand, mutant VirD2(R129G), although it retains full cleavage-and-religation capability in vitro, seems to be impaired in the putative ligation step *in planta*. The R129G mutant produces transformants at a much reduced level, and the few T-DNA inserts it produces have various small truncations at their right ends (TINLAND et al. 1995). These authors conclude that wild-type VirD2 is in some way responsible for the normally precise ligation of the right end of T-DNA to plant DNA. Because the "efficiency" of integration by VirD2(R129G), measured as the ratio of transformation events to transient expression, is about the same as that of wild type, Tinland et al. propose that the rate limiting step is integration at the left border, and they further argue that it must also be the first step, although their results do not directly address timing. Left border integration is visualized as a more or less random pairing and recombination with a spot in plant DNA with which the left border region shares some microhomology. Next, wild-type VirD2 is proposed to mediate the specificity of ligation of the right border end of T-DNA to nearby plant DNA, probably at a nick site. However, these steps could occur in reverse order while maintaining that the left border process is slower than the right.

VirE2 is required for efficient T-DNA transfer and integration, but deletion of VirE2 does not affect the "efficiency" of the specific T-DNA integration step (defined narrowly as integration events per transient expression event). Deletion of VirE2 does lead to a preponderance of T-DNA inserts exhibiting truncations at the left border yet having canonical right borders (ROSSI et al. 1996). Left end fidelity is proposed by these authors to be a result of SSB activity affording physical protection of T-strands from nucleases; alternatively SSB protein subunits may prevent interior parts of T-DNA from pairing with plant DNA, leaving only the terminal area exposed for integration. In one intriguing report, VirE2 has been shown to complement in part a RecA deletion mutant of *E. coli*, hinting at a possible direct involvement of VirE2 in the recombination process (DOMBEK and REAM 1997).

3.8 Target Sites in the Plant Genome

Analysis of the sequence of plant DNA flanking simple unique T-DNA inserts has shown that junction formation with plant DNA generally produces short deletions of host plant DNA between the points of insertion of the ends of the T-strand (MAYERHOFER et al. 1991). As discussed above, at the left end of simple T-DNA inserts, T-DNA sequence often seems to exhibit "microhomology" with short segments of target plant DNA, suggesting that the T-strand left border end may pair with target DNA during integration. Plant DNA junctions with the right border end of the T-strand usually contain precisely the predicted three nucleotides of the 25-bp border repeat.

Despite the microhomology at insertion sites, T-DNA integration is an illegitimate process (GHEYSEN et al. 1991; MAYERHOFER et al. 1991). Mapping of T-DNA inserts in various plants has demonstrated that insertion occurs at random chromosomal positions, as evidenced by the success of T-DNA gene-tagging in *Arabidopsis* (FELDMAN 1991; AZPIROZ-LEEHAN and FELDMAN 1997), but predominantly in genes (see Sect. 5.1, 6). In addition to simple T-DNA inserts, one commonly finds complex patterns of integration that consist of direct or inverted repeats of two or more T-DNAs integrated in the same locus (DE BLOCK and DEBROUWER 1991; GHEYSEN et al. 1990; JORGENSEN et al. 1987). Nucleotide sequence data indicate that plant DNA flanking T-DNA insertions can suffer deletions or rearrangements and may be duplicated to yield perfect or imperfect, direct or inverted repeats (GHEYSEN et al. 1987; MATSUMOTO et al. 1990).

3.9 Expression and Inheritance of Foreign Genes in Transgenic Plants

The ideal transformant would possess a single copy of the transgene that would segregate as a mendelian trait, and its expression would be uniform from one generation to the next. Reality falls short of this description, although reasonably ideal transformants can be found, with the difficulty depending mainly upon the plant material to be transformed, and to a lesser extent on the origin and complexity of the transgene.

Because it is common in transformation experiments to pick a few promising transformants and discard the rest, it is difficult to find data describing a whole set of transgenic plants made by one method and comparison with a set made by another method. It is the general experience of those who produce *Agrobacterium* transformants that a plant promoter attached to a reporter gene seldom if ever expresses at as high a level as the original gene from which the promoter was taken. The tissue specificity of transgene expression varies from one transformant to another (see for example, BARNES 1990; UKNES et al. 1993), and is likely not to be as selective as the original gene from which the promoter was taken (for example, UKNES et al. 1993). Because of long-lasting effects of growth of plant cells in tissue culture, the expression of the transgene in the T0 transformant is not as reliable a basis for evaluation as expression in a seed-grown T1 or T2 plant. Variable or minimal expression of the transgene may result from a growing list of phenomena. Interaction between multiple copies of the transgene can lead to cosuppression of the expression of all of them (discussed by Finer et al., this volume). This can be triggered by occurrence of an inverted repeat transgenic locus, by insertion of a transgene into a heterochromatic region leading to methylation, or by extraordinarily high expression levels leading to shutdown of expression of all copies of the transgene. Another source of variability in transgene expression is the state of the chromatin at the site of insertion, which can lead to "position effect variation". This may also have superimposed on it some epigenetic influences caused by differences in the individual cells that were transformed. One group has pointed out that considerable variability in expression is found *even among clonal plants* in the field setting (MLYNAROVA et al. 1996). There may also be position effects on variability itself. Plant breeders seek uniformity in field performance despite environmental effects, an attribute termed consistency. Clearly it does not make sense to seek to reduce position effect variation far below the level of anticipated field variation for the trait in question in plants of clonal origin. Present variability is too high, and the quest for greater consistency remains an important objective for researchers producing transgenic crop plants for agricultural use. The high hopes that scaffold attachment regions, insulators or locus control regions may eliminate such variability by shielding the transgene from surrounding chromatin remain largely unfulfilled at this writing (reviewed in BREYNE et al. 1994). Nevertheless these are research areas to watch for a breakthrough could provide substantial economy of effort.

4 Plant Genetic Engineering

4.1 Design of Vectors

A wide variety of *Agrobacterium* "disarmed" (non-tumor-inducing) vectors and vector systems have been developed. Any T-DNA vector must have left and right borders, and must either replicate in *Agrobacterium* on its own (binary vector) (DE FRAMOND et al. 1983; HOEKEMA et al. 1983) or be designed to recombine with a partner plasmid that does so (cointegrate vector) (ZAMBRYSKI et al. 1983; FRALEY et al. 1985). Binary vectors rely on two plasmids, one with the T-DNA and another with the vir genes. The *vir* gene plasmid – often called a helper – is normally a Ti plasmid from which T-DNA has been deleted. Helpers in common use include octopine-type LBA4404 (HOEKEMA et al. 1983), nopaline-type MP90 (KONCZ and SCHELL 1986) and succinamopine type EHA101 (HOOD et al. 1986). A new helper recently developed is derived from pTiChry5 (TORISKY et al. 1997), a strain virulent on soybean (BUSH and PUEPPKE 1991). Most binary vectors are derived from the wide host range plasmid RK2, for example: pBIN19 (BEVAN 1984), pGA436–439 (AN et al. 1985), and the pGPTVKan family (BECKER et al. 1992). Alternatively, the convenient set of pPZP vectors is derived from pVS1 (HAJDUKIEWICZ et al. 1994). Cointegrating vectors may recombine with artificial (ZAMBRYSKI et al. 1983) or natural (FRALEY et al. 1985) DNA on the helper Ti plasmid.

The need to use *Agrobacterium* transformation on recalcitrant plants has driven the development of higher efficiency vectors. For transfer of genes into recalcitrant plants, systems have been designed with additional copies of virulence genes or with constitutively expressing virulence genes, placed on the helper plasmid or the binary vector or a ternary plasmid (HANSEN et al. 1994). For example, the use of a binary vector with helper plasmids enhancing the production of VirG and VirE proteins can allow efficient *Agrobacterium*-mediated transfer of at least 150 kb of foreign DNA into the plant genome (HAMILTON et al. 1996). A combination cointegrating vector was developed as "superbinary" (pSB series) in which the T-DNA plasmid recombines with an intermediate vector with an extra set of *vir* genes from the supervirulent Ti plasmid A281 (see below) (KOMARI et al. 1996). Criteria for choosing a vector system are based first on what *Agrobacterium* strain has the best affinity for the plant to be transformed. The binary plasmid chosen must have a selectable marker usable in that *Agrobacterium*. The plant selectable marker, where information is available, should be one that is known to be useful in the target plant. The selectable markers available in vector systems usually have promoters designed for dicot plants; monocot transformation may require a change in gene promoter. Because T-DNA integration is quite precise at the right border, it is desirable to position the gene of interest on the right side of T-DNA and the selectable marker on the left. Such an arrangement ensures that any plant cell that receives the selectable marker gene also receives the gene of interest.

4.2 Host Range: *Agrobacterium* Strain Specificity Determinants

Agrobacterium-host range is defined by the ability of *Agrobacterium* strains to induce tumors on diverse plant genera. Wide-host-range (WHR) strains generally infect most of the 93 families of plants tested (DE CLEENE and DE LEY 1976), while strains designated limited-host-range (LHR) may infect only a few families, for example, the grape strains that infect *Vitis* and very few other plants (SZEGEDI et al. 1984). Host range is an empirical parameter that will change as technology develops. For example, many pine species formerly found resistant to *Agrobacterium* infection (DE CLEENE and DE LEY 1976) have been found sensitive to wide host range strains by more persistent testing (SEDEROFF et al. 1986; STOMP et al. 1990). Bacterial host range and virulence determinants have been attributed to differences in T-DNA genes and to structural differences in at least two *vir* loci (RISUELO et al. 1982; YANOFSKY et al. 1985). The "supervirulent" strain A281 containing pTi-Bo542, incites tumors that develop faster than those incited by isogenic strains containing other Ti plasmids, and the "supervir" trait was traced to vir genes as opposed to T-DNA genes (HOOD et al. 1986). The *virG* locus was found to be the major determinant of the "supervir" phenotype (JIN et al. 1987). The "super*virG*" gene exhibits some altered amino acid sequence in the receiver domain (portion phosphorylated by VirA) (CHEN et al. 1991). These changes may cause VirG to be more readily phosphorylated or less readily dephosphorylated.

The ability to deliver maize streak virus (MSV) to maize by T-DNA transfer, a procedure called agroinfection, has been shown to depend upon a unique characteristic of *virA* from octopine strains. The octopine-type strain A6 fails to support agroinfection of maize cells, although nopaline strain C58 does so reproducibly. This specific limitation of A6 is due to two factors: (a) VirA-$_{A6}$ is defective and can be corrected by replacing 20 amino acids with the corresponding part of VirA-$_{C58}$ (HEATH et al. 1997). (b) In addition, the chromosomal DNA of A6 produces some factor that aggravates this deficiency in VirA-$_{A6}$. If VirA-$_{A6}$ is transferred to the C58 chromosomal background, it can support agroinfection without correction. HEATH et al. (1997) speculate that in A6 but not C58, the chromosome must encode a repressor that interacts with VirA-$_{A6}$.

The necrotic response induced by WHR strains of *Agrobacterium* on grapevines has been ascribed to killing of plant cells at the site of inoculation (YANOFSKY et al. 1985). Mutations within the *virC* locus of a wide host range strain prevent the necrotic response and allow high T-DNA transfer (YANOFSKY and NESTER 1986). The role of the *vir* C product, as described in an earlier section, is to facilitate T-DNA excision by binding to "overdrive" sequences; thus a *virC* mutant would presumably be reduced in virulence (attenuated).

Vir regions from different pathogenic plasmids may vary slightly in the complement of genes they contain (HOOYKAAS et al. 1984). For example, the *virF* gene which is present in the *vir* region of the octopine Ti plasmid is absent from nopaline Ti plasmids. Mutation of *virF* leads to a weakened virulence of octopine strains on tomato (MELCHERS et al. 1990) and *Nicotiana glauca* (OTTEN et al. 1985). The *virF* gene product is active in plants. Transgenic *N. glauca* plants expressing bacterial virulence gene *virF* become hosts for nopaline strains and *virF* mutant octopine strains (REGENBURG-TUINK and HOOYKAAS 1993).

The subtle differences in the virulence genes of the different Ti plasmids make it advisable, in setting out to transform a new plant species, to determine which *Agrobacterium* strain and Ti plasmid are best adapted to the plant. Likewise it is advisable to screen genotypes of the plant in order to identify the most promising plant material for transformation. The role of the plant in *Agrobacterium*-host compatibility is certainly of as much importance as that of the pathogen. For example, BYRNE et al. (1987) identified transformable soybean germplasm by such a screen.

4.3 Tissue Requirements

Ability to cultivate plant tissue in vitro is a prerequisite in almost all current transformation protocols. Transformation requires competent (i.e., transformable) cultured cells that are embryogenic or organogenic. Plant cells suitable for regeneration are cocultivated with *Agrobacterium*, followed by a procedure to select and regenerate transformed cells/tissues at a reasonable frequency. As a point of departure, it is advisable to use transient expression of reporter genes to identify conditions that allow T-DNA transfer. Next one must find conditions for gene-transfer into the largest possible number of plant cells without affecting tissue survival, regenerability, and ultimately fertility.

Transient expression studies are generally performed with bacterial *uid*A (GUS) gene encoding a β-glucuronidase carried by the T-DNA (JEFFERSON 1987). A caveat in such studies is that marker expression can be the handiwork of contaminating *Agrobacteria*, which have an uncanny ability to read "plant-specific" gene promoters. To circumvent this source of background the prokaryotic ribosome binding site sequence immediately upstream of the GUS coding sequence of the *uiad*A gene has been modified (JANSSEN and GARDNER 1989). Further, an intron has been inserted in the coding sequencing of the gene (VANCANNEYT et al. 1990). Alternatively an intron with a termination codon in the same reading frame as the GUS coding sequence has been inserted within the 5' end of the gene, such that removal of the intron is required for expression (OHTA et al. 1990). ROSSI et al. (1993) achieved GUS expression exclusively in the plant cells by use of a translational start site derived from gene V of cauliflower mosaic virus.

4.4 General Methods of Transformation

A leaf disc transformation protocol (FRALEY et al. 1983; HORSCH et al. 1985) is generally useful for many dicots. Surface-sterilized leaf discs or cotyledons are inoculated with *Agrobacterium* carrying a selectable marker on a disarmed vector. Explants are then subcultured for some days on medium favoring shoot formation. Subsequently tissues are transferred to the same medium containing appropriate antibiotics to kill the bacteria and the agent to select for transformants. Surviving tissues are then rooted and transferred to soil. In general, selection is applied after a cocultivation period of 1–5 days. Cocultivation of *Agrobacterium* with stem segments, suspension cultures, microcalli and germinating seeds has proven to be similarly successful. The majority of the experiments have been performed with *A. tumefaciens*, but some employed *A. rhizogenes*. Many variables can be altered, such as: *Agrobacterium* strain, use of "feeder cells", pH, vacuum infiltration, osmoticum treatment, duration of cocultivation, antioxidants, sonication, biolistic wounding prior to infection with *Agrobacterium*, and temperature at the time of cocultivation (BYRNE et al. 1987; BIDNEY et al. 1992; PERL et al. 1996; TRICK and FINER 1998; DILLEN et al. 1997; FULLNER and NESTER 1996; FULLNER et al. 1996). Contrary to folklore, it appears that wounding of the plant tissue prior to inoculation is not absolutely essential for *Agrobacterium*-mediated transformation: *Agrobacterium* can use stomatal openings to access target cells (ESCUDERO and HOHN 1997).

One very simple transformation system differs from the protocol described above in that it avoids the use of tissue culture. It involves infiltration of *Agrobacterium* cells into *Arabidopsis* plants before flowering, and direct selection for transformants among the progeny plants (FELDMAN 1991; BECHTOLD et al. 1993; BENT and CLOUGH 1998). In one study the T-DNA copy number was low, averaging 1.4 inserts per transformant, but T-DNA inserts were found to consist of complex concatemers of direct and inverted repeats (FELDMAN 1991), a situation likely to produce low transgene expression (HOBBS et al. 1993) and at least one form of gene silencing (IGLESIAS et al. 1997). Although this system has been developed and refined for gene tagging (maximizing the number of hits) rather than engineering the plant, one should be able to sacrifice "efficiency" if necessary and adapt this approach to produce simple patterns of T-DNA insertion. Any transformation procedure that avoids tissue culture can potentially circumvent such problems as aneuploidy, sterility and epigenetic differences among target cells. It will be interesting to see how widely this approach can be applied to different types of plants.

4.5 Transformation of Recalcitrant Plants

Some crops react to *Agrobacterium* infection by necrosis. In general, phytopathogenic bacteria carry *hrp* (hypersensitive reaction and pathogenicity) genes whose products are involved in the elicitation of the hypersensitive response (HR; necrosis) in resistant plants (BAKER et al. 1997). Interestingly, *Agrobacterium* has been reported not to elicit a typical hypersensitive response in most plants (NAM et al. 1997). Tissue necrosis involving T-DNA genes is observed in certain plant-*Agrobacterium* interactions (DENG et al. 1995), but the symptoms appear more slowly than in a typical HR and necrosis extends beyond the inoculation site. Very short exposure of embryogenic calli of grape plants (*Vitis vinifera*) to diluted cultures of *Agrobacterium* resulted in plant tissue necrosis and subsequent death. Killing seemed to be oxygen-dependent and correlated with elevated levels of peroxides. Treatments with antioxidants such as polyvinylpyrrolidone and dithiothreitol can improve plant survival and enable the recovery of stable transgenic grape plants (PERL et al. 1996).

Necrosis in shoot tips of both aspen and poplar has been overcome by addition of Ca-gluconate and buffering the medium with 2-(*N*-morpholino)ethanesulfonic acid, and by growing the shoots below 25°C. Necrosis of the explants was probably due to an accumulation of ammonium in the cells and was overcome by adjusting the NO_3^-/NH_4^+ ratio of the medium (DE BLOCK 1990).

Monocotyledonous plants have seemed refractory to *Agrobacterium*-mediated transformation, but the reason for this recalcitrance was unknown. Maize was probably the most thoroughly studied system. In maize it was demonstrated that the *vir* genes were induced and T-DNA transfer and nuclear targeting occurred, as shown by transient assays on maize tissues and by agroinfection experiments (GRIMSLEY et al. 1989; SHEN et al. 1993; ZUPAN et al. 1996). Moreover VirD2, whatever its role may be in the specificity of T-DNA integration into chromosomes of dicot plant cells, seems to function similarly in maize as demonstrated by the success of Agrolistic transformation (discussed below) (HANSEN et al. 1997a). Transformation could be hampered by toxic maize compounds which reduce the *Agrobacterium* population or specifically interfere with *vir* gene induction (SAHI et al. 1990). A block in T-DNA integration was suggested to underlie the recalcitrance of maize, and possibly other monocots (NARASIMHULU et al. 1996). The same hypothesis was proposed on the basis of the lack of cell division during monocot wound response (BINNS and THOMASHOW 1988). Nevertheless, efficient *Agrobacterium*-mediated transformation of various tissues from rice (HIEI et al. 1994) as well as of maize embryos from the inbred A188 and its derived hybrids has been reported (ISHIDA et al. 1996). The use of a "super-binary" vector containing an extra set of virulence genes (*virG*, *virC* and *virB*) from the supervirulent strain A281 appeared to contribute to this accomplishment. The nature of the problem and the identity of the solution remain obscure, but maize and other monocots have now joined the list of reproducibly transformable plants.

4.6 Transgenic Plants Free from Selectable Marker Genes

Transformation systems that have been developed for the introduction of foreign genes into plant cells almost always include genes that confer a selectable advantage on transformed cells. These genes are not necessary once the transgenic plants are produced and indeed the removal of the selectable marker gene may be desirable for several reasons: (a) The presence of a functional selectable marker precludes the use of the same selection system in subsequent transformations of derivatives of that line. (b) For food crops there is some apprehension that the products formed from a herbicide or other selection agent might be toxic or allergenic when ingested. (c) For the case of antibiotic resistance markers a concern has been raised that such drug resistance could be transferred to enteric micro-organisms, increasing the possibility of antibiotic-resistant pathogenic micro-organisms in the human or animal gut. (d) If the marker is a herbicide resistance gene, there is concern that such crops may lead to increased herbicide use, or (contrariwise) that the herbicide resistance gene may escape by wide crossing into weeds closely related to the crop, thus rendering the herbicide less useful.

YODER and GOLDSBROUGH (1994) have reviewed several approaches for elimination of selectable markers. Cotransformation is one method for producing transgenic plants free from the selectable marker. In this system the marker gene and the gene of interest are placed on separate T-DNAs, either in the same strain or in two different *Agrobacterium* strains. The separate T-DNAs can integrate into the plant genome at different locations. Consequently the gene of interest can segregate from the marker gene in the progeny. Cotransformation has been tested in several plant systems. In *Brassica napus*, using two *Agrobacterium* strains and nopaline Ti-derived vectors DE BLOCK and DEBROUWER (1991) found 39%–85% cotransform-ants, of which 78% had *linked* T-DNAs, usually inverted repeats joined at their right borders. In contrast, cotransformation of tobacco leaf discs with two *Agrobacterium* strains carrying different pRK290-derived binary vectors in octopine-type LBA4404 produced 11 KmR plants, of which only three carried the second marker (nopaline synthase) (McKNIGHT et al. 1987). Analysis of outcrossed progeny of the three double transformants showed independent segregation of Nos and KmR traits, showing that the traits were integrated at *unlinked* sites in each case (McKNIGHT et al. 1987). These differences could reflect differences between nopaline Ti and octopine Ti plasmids in tendency toward formation and/or inte-gration of T-DNA concatemers. However, other differences could be crucial: higher copy number of the binary vector in McKnight's strain, or the plants studied (tobacco vs. brassica) or the tissue (leaf disks vs. hypocotyl explants) or the methodology (single vs. double selection). In a further study on the role of the plant and tissue in determining the course of events, DE NEVE et al. (1997) tested *Arab-idopsis* leaf discs vs. *Arabidopsis* roots vs. tobacco protoplasts for frequencies of linked vs. unlinked cotransforming inserts. The *Arabidopsis* systems integrated mainly single copies of the two T-DNAs, with ca. 50%–75% linked cotransform-ants. Tobacco protoplasts produced more multicopy events: of the ca. 50% single copy transformants, about half exhibited linkage between the cotransformed T-DNAs. These studies taken together argue that in dicots segregating cotransformed T-DNAs can usually be found, but often not at high frequency.

Delivery of two separate T-DNAs from a single *Agrobacterium* (DEPICKER et al. 1985; KOMARI et al. 1996) gives higher cotransformation frequency as one might expect, with reasonable segregation frequency. The superbinary system developed by Komari for monocot transformation appears to favor unlinked integration sites. For example, in rice, KOMARI et al. (1996) observed 47% co-transfer of separate T-DNAs from the same *Agrobacterium*, and 65% of these allowed segregation of the selectable marker from a GUS test gene. In tobacco the same vector system gave 52% cotransfer and 55% segregation (KOMARI et al. 1996).

Another approach to elimination of the selectable marker gene is to flank the gene with direct repeats of recognition sites for a site-specific recombinase, which allow the enzyme to excise the marker. Using the Cre/*lox* system, DALE and OW (1991) created a recipient plant with a Luciferase (Luc) gene and a HygR (Hygromycin) marker, with the latter flanked by *lox* sites. They then retransformed with a 35S-Cre-nos KmR construct and regenerated 11 KmR plants, of which 10 had lost HygR. They achieved the same result by crossing the LOX-HygR-LOX plant to a 35S-Cre-nos plant, obtaining 42/78 Luc+ plants that appeared HygS. Similar experiments by RUSSEL et al. (1992) demonstrated that the introduction of 35S-Cre-nos gene through a genetic-cross produced T2 plants that were mosaic for the excision event, with pure excision lines appearing at the T3 generation. Such a delay poses no problem in view of the time needed to segregate the introduced recombinase from the gene of interest. For eliminating the selectable marker, the recombinase excision approach has an advantage in efficiency over the two T-DNA approach described above: a single T-DNA carrying gene of interest plus the selectable marker can be employed, giving 100% cotransfer if the selectable marker is positioned at the left border. However, it is essential that both recognition sites for the recombinase be integrated, an argument for situating the selectable marker at the *right* border in this case and screening for complete transfer of the gene of interest.

4.7 Targeted Integration Into the Plant Genome

Because of position effect variation in the expression level and pattern of transgene expression, it would be desirable to be able to choose in advance a target site into which the transgenes will be directed. This could be achieved either by using site-specific recombinase technology or by homologous recombination into the plant genome. While the preferred mode of *Agrobacterium* T-DNA insertion is clearly illegitimate recombination, nevertheless efforts toward coercing this vector to undergo homologous recombination have been gallant. Most experiments have used an artificial transgenic target site containing only part of a selectable marker, requiring precise recombination with donor DNA to produce a functional gene. Some have exerted strong selection for loss of target site DNA. In nearly every case the recipient plant finds ways to foil the selection scheme with false positive transformants of various kinds (for example, see THYKJAER et al. 1997). We note that the term "homologous" has been used to embrace several types of T-DNA transformation events, including:

A. Gene conversion at the target locus. The defective target site marker is corrected by copying missing information from donor DNA (before or after illegitimate integration elsewhere – possibly nearby – in the plant genome).

B. Homologous recombination with the target locus on one end of T-DNA and illegitimate insertion at the other.

C. Double crossover that replaces target DNA by donor DNA through precise homologous recombination on both sides of the target locus.

The type C event produces what the genetic engineer needs: an entirely predictable product with no unforeseen deletions. However, any of the three processes would suffice to "tag" the gene, which is often the objective of investigators pursuing this approach. Given a DNA clone, they want to create a plant mutant at that locus and determine the phenotype. Distinguishing A, B, and C type insertion events requires PCR, Southern blots, and analysis of progeny. Type C events should produce homozygous progeny that lack the original target site.

For homologous recombination of T-DNA into the tobacco genome, all successful experiments have employed donor DNA that carries only a part of a selectable marker, relying on the target site to complete it. For example, a part of the sulfonylurea herbicide resistant AHAS marker yielded three type A recombinants (Lee et al. 1990). In the hands of the Hooykaas group, a piece of the NPTII (KanR) marker produced one recombination (type unclear) for Offringa et al. (1990), one type A (Offringa et al. 1993), and four probably type B plus one type C (Risseeuw et al. 1995) recombination events. This group also demonstrated that the use of an intact marker gene together with powerful negative selection for loss of the target locus is *not* a good approach (Risseeuw et al. 1997): small deletions are surprisingly common in the genome of protoplast-derived tobacco cells, and all of the KmR clones selected by this approach had illegitimate inserts and deletions of the target site. A double-strand break in genomic DNA greatly enhances the probability of homologous recombination that repairs the break. Using a genomic I-*Sce*I cut site as target, Puchta et al. (1996) raised the frequency of homologous integration events to ca. 1% of illegitimate ones, with "type C" events constituting about half of all targeted ones (Puchta et al. 1996).

Efforts to achieve homologous recombination into the (much smaller) genome of *Arabidopsis* have predictably been more successful. Miao and Lam (1995) demonstrated the power of a PCR screen on pools of 10 *Agrobacterium*-mediated transformants made with an intact KmR selectable marker. They found one Type C homologous event out of 2580 transformants screened, but the cultures transformed were not regenerable. This group (Kempin et al. 1997) has reported a second type C disruption, this time in regenerable material. The homozygous progeny *Arabidopsis* plants were demonstrated to be free from transcript of the targeted locus, good evidence that it had indeed been "knocked out".

Although a PCR screen on pools of 10 *Arabidopsis* transformants is feasible, on a "real" plant with 20- to 50-fold larger genome size, the screen might have to be done plant by plant. The ratio of homologous events to illegitimate insertions might be similar for T-DNA transformation, or it might be reduced by the factor of the genome size. It probably does not make sense to compare the numbers from all of these publications because the length of homology flanking the expected recombination region varied. If we do so nevertheless, we see a frequency of around 10^{-4}–3×10^{-6} compared to illegitimate insertion events. This is similar to the ratio of homologous to illegitimate insertions using direct gene transfer into protoplasts (see, for example, PASZKOWSKI et al. 1988; HALFTER et al. 1992), a topic outside the scope of this review.

From all of the studies summarized above it is clear that finding homologous double-recombination (type C) products among T-DNA transformants is very far from being a routine, or even feasible, procedure at this time, despite many clever approaches. Several recent discoveries may raise new hope for the feasibility of homologous integration. The moss *Physcomitrella patens*, whose haploid phase is the predominant part of its life cycle, exhibits homologous integration at high frequency. Perhaps this moss can teach us its secrets and allow us to apply them to higher plants (SCHAEFER and ZRYD 1997). In addition, the analysis of radiation sensitive mutant *Arabidopsis* lines that are impaired in T-DNA integration (NAM et al. 1997; SONTI et al. 1995) should be useful in analyzing both homologous and illegitimate recombination processes and the genes involved. The mutants studied by Sonti et al. (1995) have been demonstrated to be transformable by *Agrobacterium* using other methods (MYSORE et al. 1988); nevertheless, the mutations are likely in a step contributing to insertion. Such knowledge may suggest new ways to persuade plants to integrate   T-DNA at homologous sites.

Site-specific recombinases, which have been used with some success to target DNA delivered to plant protoplasts, require double-stranded DNA as substrate and might appear to be a poor choice for targeting T-DNA, a single-stranded DNA coated with proteins. Nevertheless, VERGUNST et al. (1998) have achieved dramatic success using the Cre/*lox* system to target a KmR construct lacking promoter and ATG start codon to the missing components inserted into the *Arabidopsis* genome. In the target site, the promoter/initiation codon to be captured by the incoming construct is exploited in the meantime to drive the *cre* gene and produce recombinase, which becomes undesirable after the insertion because it can catalyze the reverse reaction and reexcise the donor DNA. Plants hemizygous for the target site produced correct inserts in 39 of 44 KmR calli selected. The frequency of these events was ca. 1% that of random T-DNA insertions measured separately in the same material. It is not clear whether the substrate for recombinase excision of the donor construct in this experiment was a double-stranded form of T-DNA before or during random integration, or whether random T-DNA integration elsewhere precedes excision and site-specific integration. ALBERT et al. (1995) have used a similar strategy to achieve site-specific integration of naked DNA into tobacco protoplasts, but the donor DNA *lox* had to be a mutant site in order to prevent reexcision. Alternatively they succeeded with wild-type lox sites when Cre was produced by transient expression. In both cases, only about half of the selected events were correctly targeted, far below the efficiency of the *Agrobacterium*-mediated Cre/*lox* targeted insertion system developed by VERGUNST et al. (1998).

5 Application of Plant Genetic Engineering

The *Agrobacterium* system is being extensively used for the transfer of various traits to plants as well as for the study of gene function in plants. Applications include the transfer of genes affecting: resistance to virus, herbicide tolerance, altered shelf life of tomato, male sterility, resistance to pathogenic bacteria (HOOYKAAS and SCHIL-PEROORT 1992). In addition this microbe has aided in identification of gene parts useful for genetic engineering. In the most direct and complete utilization, *Agrobacterium* can colonize transgenic plants and has shown us how to create an artificial symbiotic relationship between an engineered plant and a protective microbe.

5.1 Mutagenesis and Gene Traps

T-DNA is a powerful tool for induction of insertional gene mutations in plants. The *Agrobacterium* system has also been used to tag and therefore to identify plant genes influencing plant development. This approach has been especially successful for *Arabidopsis thaliana* for which large numbers of independent T-DNA-tagged mutants have been obtained (FELDMANN 1991). Also some "promoter trap" T-DNA vectors have been developed that have a promoterless resistance or scorable marker gene situated just inside the right border. Activation of expression can occur after integration into a transcriptionally active area. It has been found that such gene activation occurs at high frequency (30%–50%) suggesting that the T-DNA integrates preferentially in transcriptionally active areas. These vectors, named gene traps and enhancer traps according to their design, are reporter genes that are not normally expressed unless they are integrated near or within a chromosomal gene. Enhancer traps are equipped with a minimal promoter that can respond to nearby enhancers; gene traps are equipped with a splice acceptor so that integration within introns leads to readthrough transcription and splicing. In each case the expression of the resulting chimeric reporter gene closely mimics that of the chromosomal gene (BABIYCHUK et al. 1997; SUNDARESAN et al. 1995; MARTIENSSEN 1998). Large collections of lines have been generated and used extensively for both developmental biology and genomic research. Such enhancer and gene trap vectors can also be combined with mobile elements.

5.2 Promoters, Selectable Markers and Hormone-Synthesizing Genes

Among the different bacteria that can infect plants, *Agrobacterium* is unique in the sense that it harbors in its T-DNA genes with eukaryotic transcription signals. Whether this trait is a fossil of an ancient genetic exchange system or a parallel evolution of DNA sequences that provide selective advantage in plant-bacterial interaction is an intriguing question (for review see OTTEN et al. 1992). Many T-DNA promoters have been used to create functional chimeric genes, for example, those controlling the T-DNA genes octopine synthase and nopaline synthase or some oncogenes (HANSEN et al. 1997b) and more recently the dual promoter of the mannopine synthase genes, pmas (BECK VON BODMAN et al. 1995). Enhancerlike properties that have been ascribed to the ocs promoter are used (NI et al. 1995) in a super-*mas* (mannopine synthase) promoter. Some oncogenes (onc) are responsible for the synthesis of hormones in plants. They have thus contributed to our understanding of hormone function in plants (SMIGOCKI and OWENS 1988; ESTRUCH et al. 1991). Hormone biosynthesis has also proven to be useful as a negative selectable marker (CZAKO et al. 1996).

5.3 Artificial Symbiosis

The opine concept has played an important role in our view of the ecological niche provided by the crown gall tumor for the bacterial pathogen. The tumor is instructed by T-DNA to make precisely the opines that the pathogenic strain can utilize. Further, the gall produces at least one conjugative opine, which is the inducer for Ti plasmid transfer from one *Agrobacterium* strain to another (PETIT et al. 1978). The central player appears to be the Ti plasmid, which as a result of the gall begets more Ti plasmid copies through growth of its host and propagates itself to additional hosts in the setting of the gall.

One direct application of the opine concept is the engineering of normal plants producing opines in order to favor growth of a target population of bacteria colonizing the rhizosphere. The opine thus establishes a trophic link between the two partners. If the bacteria are potentially beneficial to the plant, this association may turn into an artificial symbiosis, with the micro-organism protecting the plant that feeds it (LAM et al. 1997; OGER et al. 1997; SAVKA and FARRAND 1997).

5.4 T-DNA Insertion without *Agrobacterium*:
Agrolistic Transformation

We have devised an extremely simple method for human T-DNA delivery to plants, based on the generation of T-strands *in planta* (HANSEN and CHILTON 1996; HANSEN et al. 1997a). Using the Biolistic device, plasmid DNA with plant-expressible forms of *virD1* and *virD2* was codelivered to tobacco cells together with a substrate plasmid carrying a selectable marker and gene of interest flanked by left and right borders. Transient expression of VirD1 and VirD2 proteins generates T-strands *in planta* that insert into the plant genome in their usual manner. T-DNA/plant DNA junctions cloned from the transformed plant cells have precisely the 3 bp of the right border repeat expected for a T-DNA insert. The method of delivery of DNA is not an essential aspect for agrolistic transformation: maize protoplast transformation by these DNAs likewise yields T-strand inserts (HANSEN et al. 1997a). Unlike *Agrobacterium* T-DNA transformation, the agrolistic process does not absolutely require VirE2, although inclusion of a plant-expressible *virE2* plasmid does increase the number of transformants somewhat. In agrolistic transformation, VirD1 and VirD2 are presumably produced in the plant cytoplasm, but where they generate T-strands is unknown. VirD2 can enter the nucleus efficiently, but VirD1, which is not thought to enter the plant cell in normal *Agrobacterium* T-DNA delivery, has not to our knowledge about been tested for NLSs. The finding that *Agrobacterium* can deliver free VirE2 to the plant suggests the possibility that free VirD2, separate from T-strands, might also be delivered to the plant by *Agrobacterium*. It is interesting to speculate what possible role it might play in the integration process. The Agrolistic procedure, developed initially as a method for transforming recalcitrant plants, may lead us to new insights into the intricate methodology developed by *Agrobacterium*.

6 Limitations of T-DNA as a Gene Vector

There are recent reports that *Agrobacterium* surprisingly often transfers the wrong side of the plasmid – the vector backbone – to the plant by cotransformation with T-DNA (RAMANATHAN and VELUTHAMBI 1995; KONONOV et al. 1997; WENCK et al. 1997). In one case 75% of transgenic tobacco plants contained vector DNA (KONONOV et al. 1997). This unwanted DNA was either transferred to the plant cell independently of T-DNA or linked to the T-DNA across either the left or the right T-DNA border. These phenomena could result from the initiation of T-strand formation at the left border, or by skipping the left border in T-strand production. Although there is no selection pressure for the transfer of such sequences, we have seen that cotransfer of unselected T-DNAs can be quite efficient. As discussed in an earlier section, we have little understanding of how and when the left border of T-DNA is cut to terminate the T-strand correctly. Research in this area should provide information valuable in addressing this problem. It may be that in the course of developing high efficiency vectors that produce ever-higher numbers T-strands, we have actually increased this type of error. Overexpression of VirD1 and VirD2 in superbinary strains may lead to unemployed protein molecules that overzealously attack left borders and initiate V-strands (vector DNA) rather than T-strands. Some vectors use synthetic 25-bp border sequences rather than long fragments of natural border context from the Ti plasmid; they may thus lack some of the characteristics that allow VirD1/VirD2/VirC to identify the right border and left border correctly.

A second problem of T-DNA insertions that is not unique to this method of plant transformation is the interruption of a plant gene at the target site of insertion. From experiments using T-DNA as a promoter trap, it is clear that T-DNA inserts preferentially into genes (KONCZ et al. 1989) based on the finding that 30% of inserts create gene fusions. This value is the same for tobacco as for *Arabidopsis*, despite the 20-fold larger genome size of tobacco. It is clear therefore that the hits are not random, and in fact the bias is toward interrupting genes. This problem is the dark side of the great utility of T-DNA insertions for gene-tagging: the mutations caused by T-DNA insertions have formed a library of tagged genes for *Arabidopsis* (FELDMAN 1991), and this could apparently be done for additional plants. However, such mutations are a potential problem for the genetic engineer. Especially bothersome is the fact that the problem does not become apparent until the plant is selfed: the damage, if any, is recessive. Redundancy of functions in the plant genome may minimize the impact of many hits. Site-specific insertion of DNA into a specified target site in the plant genome (possibly repaired when it was identified) would ameliorate this situation. It is clearly a potential problem, whose magnitude we cannot foresee at present.

7 Conclusion

Our understanding of the details of *Agrobacterium* T-DNA transfer to plants has come a remarkable distance. We certainly know enough to domesticate this natural delivery system to our own purposes. It is now clear how to build vector plasmids of increasing utility, and the plant literature is a rich source of selectable marker genes, introns, terminators, enhancers, tissue-specific and constitutive promoters, which can be linked together with genes of interest in crop protection as they are discovered and developed. Some of the pitfalls of vector design are easily circumvented, e.g., the use of natural borders and context DNA to assure that VirD1/VirD2 will understand what we intend to be right and left borders. It is clear that an abundance of VirE2 is needed in the plant cell for protection of T-strands, maximizing chances of delivery and insertion of full-length T-DNA. We know that the selectable marker should be situated leftward of the genes of interest in order to assure that they are incorporated intact. In short, we know the mechanics of vector design fairly well.

Many of the challenges that remain are still of considerable importance to users of this vector system. We do not understand the process of finishing the T-strand at the left border, or why vector DNA is included in some cases. We have little understanding of how T-strands "choose" their sites of insertion in the plant genome, although insertion sites are grossly random. Why do we find microhomology on the left end? T-strands preferentially are in genes, but how does the T-strand recognize genes from filler DNA in the plant genome? Are transcribed regions favored? Are methylated heterochromatic regions avoided? We do not know whether it is better for *Agrobacterium* to overexpress *vir* genes and make a large stockpile of T-strands, or whether that will lead to incorporation of multiple inserts and gene silencing. We do not know how to control the copy number of inserts, or whether one *Agrobacterium* strain is better than another for the purpose of making simple inserts. We do not understand how T-strands are linked together in direct and inverted orientation – whether this occurs before or after insertion of an initial copy into plant DNA.

Basic science interest remains in the area of structures. The assembly of the numerous proteins of the *virB* operon into a bridge from *Agrobacterium* to the plant cell is being intensively studied. The place of the newly discovered pili in this process will be interesting to follow. Studies of the chemistry of *Agrobacterium* binding to the plant cell will bring new insights, and may explain some examples of host range limitation.

We therefore conclude with the confidence that *Agrobacterium* as a research topic is still alive and well, and promises to yield new insights for many years to come. It is our hope that the mechanism of *Agrobacterium* T-DNA transfer may prove a source of continuing interest to scientists in the many disciplines to which this clever bacterium has made such varied and valuable contributions.

Acknowledgements. The authors thank Dr. Barbara Hohn and Dr. Stan Gelvin for valuable suggestions on the manuscript and for permission to cite unpublished results.

References

Albert H, Dale EC, Lee E, Ow DW (1995) Site-specific integration of DNA into wild-type and mutant *lox* sites placed in the plant genome. Plant J 7(4):649–659

Albright LM, Yanofsky MF, Leroux B, Ma D, Nester EW (1987) Processing of the T-DNA of *Agrobacterium tumefaciens* generates border nicks and linear, single-stranded T-DNA. J Bacteriol 16:1046–1055

Alt-Moerbe J, Rak B, Schroder (1986) A 3.6-kbp segment from the *vir* region of Ti plasmids contains genes responsible for border sequence-directed production of T region circles in *E. coli*. EMBO J 5:1129–1135

An G, Watson BD, Stachel S, Gordon MP, Nester EW (1985) New cloning vehicles for transformation of higher plants. EMBO J 4:277–284

Azpiroz-Leehan R, Feldmann KA (1997) T-DNA insertion mutagenesis in Arabidopsis: going back and forth. TIG 13:152–156

Babiychuk E, Fuangthong M, Van Montagu M, Inzé D, Kushnir S (1997) Efficient gene tagging in *Arabidopsis thaliana* using a gene trap approach. Proc Natl Acad Sci USA 94:12722–12727

Baker B, Zambryski P, Staskawicz B, Dinesh-Kumar S P (1997) Signaling in plant microbe interactions. Science 276:726–733

Ballas N, Citovsky V (1997) Nuclear localization signal binding protein from Arabidopsis mediates nuclear import of Agrobacterium VirD2 protein. Proc Natl Acad Sci USA 94:10723–10728

Barnes WM (1990) Variable patterns of expression of luciferase in transgenic tobacco leaves. Proc Natl Acad Sci USA 87:9183–9187

Baron C, Llosa M, Zhou S, Zambryski PC (1997a) VirB1, a component of the T-complex transfer machinery of *Agrobacterium tumefaciens*, is processed to a C-terminal secreted product, VirB1*. J Bacteriol 179:1203–1210

Baron C, Thorstenson YR, Zambryski PC (1997b) The lipoprotein VirB7 interacts with VirB9 in the membranes of *Agrobacterium tumefaciens*. J Bacteriol 179:1211–1218

Bechtold N, Ellis J, Pelletier G (1993) In planta Agrobacterium mediated gene transfer by infiltration of adult *Arabidopsis thaliana* plants. C R Acad Sci Paris 316:1194–1199

Beck von Bodman S, Domier LL, Farrand SK (1995) Expression of multiple eukaryotic genes from a single promoter in Nicotiana. Bio/Technology 13:587591

Becker D, Kemper E, Schell J, Masterson R (1992) New plant binary vectors with selectable markers located proximal to the left T-DNA border. Plant Mol Biol 20:1195–1197

Bent AF, Clough SJ (1998) Agrobacterium germ line transformation: transformation of Arabidopsis without tissue culture. In: Plant Molecular Biology Manual B7, Kluwer Academic, The Netherlands

Berger BR, Christie PJ (1994) Genetic complementation analysis of the *Agrobacterium tumefaciens* virB operon: virB2 through virB11 are essential virulence genes. J Bacteriol 177:4890–4899

Bevan, M (1984) Binary Agrobacterium vectors for plant transformation. Nucleic Acids Res 12:8711–8721

Bevan MW, Chilton M-D (1982) T-DNA of the Agrobacterium Ti and Ri plasmids. Ann Rev Genet 16:357–384

Bidney D, Scelonge C, Martich J, Burrus M, Sims L, Huffman (1992) Microprojectile bombardment of plant tissues increases transformation frequency by *Agrobacterium tumefaciens*. Plant Mol Biol 18:301–313

Binns AN, Thomashow MF (1988) Cell biology of Agrobacterium infection and transformation of plants. Ann Rev Microbiol 42:575–606

Binns AN, Beaupré CE, Dale EM (1995) Inhibition of VirB-mediated transfer of diverse substrates from *Agrobacterium tumefaciens* by the IncQ plasmid RSF1010. J Bacteriol 177:4890–4899

Bravo-Angel AM, Hohn B, Tinland B (1998) The omega sequence of VirD2 is important but not essential for efficient transfer of T-DNA by *Agrobacterium tumefaciens*. Mol Plant-Microbe Interact. 11:57–63

Breyne P, Van Montagu M, Gheysen G (1994) The role of scaffold attachment regions in the structural and functional organization of plant chromatin. Transgenic Res 3:195–202

Bush AL, Pueppke SG (1991) Characterization of an unusual new *Agrobacterium tumefaciens* strain from *Chrysanthemum moriflorium* Ram. Appl Environm Microbiol 57:2468–2472

Byrne MC, Mcdonnell R, Wright MS, Carnes MC (1987) Strain and cultivar specificity in the Agrobacterium-soybean interaction. Plant Cell Tissue Organ Culture 8:3–15

Cangelosi GA, Martinetti G, Leigh JA, Lee CC, Theines C, Nester EW (1989) Role of *Agrobacterium tumefaciens* ChvA protein in export of β-1,2 glucan. J Bact 171:1609–1615

Chen CY, Wang L, Winans SC (1991) Characterization of the supervirulent virG gene of the *Agrobacterium tumefaciens* plasmid pTiBo54. Mol Gen Genet 230:302–309

Chilton M-D, Drummond MH, Merlo DJ, Sciaky D, Montoya AL, Gordon MP, Nester EW (1977) Stable incorporation of plasmid DNA into higher plant cells: the molecular basis of crown gall-tumorigenesis. Cell 11:263–271

Chilton M-D, Tepfer DA, Petit A, David C, Casse-Delbart F, Tempe J (1982) *Agrobacterium rhizogenes* inserts T-DNA into the genomes of the host plant root cells. Nature (Lond) 295:432–434

Christie PJ (1997) *Agrobacterium tumefaciens* T-complex transport apparatus: a paradigm for a new family of multifunctional transporters in eubacteria. J Bacteriol 179:3085–3094

Christie PJ, Ward JE, Gordon MP, Nester EW (1989) A gene required for transfer of T-DNA to plants encodes an ATPase with autophosphorylating activity. Proc Natl Acad Sci USA 86:9677–9681

Citovsky V, De Vos G, Zambryski P (1988) Single-stranded DNA binding protein encoded by the virE locus of *Agrobacterium tumefaciens*. Science 240:501–540

Citovsky V, Zupan J, Warnik D, Zambryski P (1992) Nuclear localization of Agrobacterium VirE2 protein in plant cells. Science 256:1802–1804

Czako M, Wenck AR, Marton L (1996) Negative Selection markers for plants. In: Gresshoff P (ed) ACRC series of current topics in plant molecular biology. Technology transfer of plant biotechnology, vol 4. CRC, Boca Raton, pp 67–93

Dale EC, Ow DW (1991) Gene transfer with subsequent removal of the selection gene from the host genome. Proc Natl Acad Sci USA 88:10558–10562

Das A (1994) Regulation of *Agrobacterium tumefaciens* virulence gene expression. In Molecular Mechanisms of Bacteria Virulence. Kado CI, Crosa JH Kluwer Academic Publishers. Netherlands, 477–489

De Block M (1990) Factors influencing the tissue culture and the *Agrobacterium tumefaciens*-mediated transformation of hybrid aspen and poplar clones. Plant Physiol 93:1110–1116

De Block M, Debrouwer D (1991) Two T-DNAs co-transformed into Brassica napus by a double Agrobacterium infection are mainly integrated at the same locus. Theor. Appl Genet 82:257–263

De Cleene M, De Ley J, (1976) The host range of crown gall. Botanical Review 42:389–466

De Framond AJ, Barton KA, Chilton MD (1983) Mini-Ti: a new vector strategy for plant genetic engineering. Bio/Technology 1:262–269

De Neve M, De Buck S, Jacobs A, Van Montagu M, Depicker A (1997) T-DNA integration patterns in co-transformed plant cells suggest that T-DNA repeats originate from co-integration of separate T-DNAs. Plant J 11:15–29

Deng W, Pu X, Goodman R, Gordon M, Nester E (1995) T-DNA genes responsible for inducing a necrotic response on grape vines. Mol Plant-Microbe Interact. 8:538–548

Deng W, Chen L, Wood DW, Metcalfe T, Liang X, Gordon MP, Comai L, Nester EW (1998) Agrobacterium VirD2 protein interacts with plant cyclophilins. Proc Natl Acad Sci 95:7040–7045

Depicker A, Herman L, Jacobs A, Schell J, van Montagu M (1985) Frequencies of simultaneous transformation with different T-DNAs and their relevance to the Agrobacterium plant cell interaction. Mol Gen Genet 201:477–484

Dessaux Y, Petit A, Tempé J (1991) Opines in Agrobacterium biology. In: Verma DPS (ed) Molecular signals in plant-microbe communication. CRC, Boca Raton, pp 109–136

De Vos G, Zambryski, P (1989) Expression of Agrobacterium nopaline specific VirD 1, VirD2 and VirC1 and their requirement for T-strand production in E. coli. Mol Plant Microbe Inter. 2:43–52

Dillen W, De Clercq J, Kapila J, Zambre M, Van Montagu M, Angenon G (1997) The effect of temperature on Agrobacterium tumefaciens-mediated gene transfer to plants. Plant J 12:1459–1463

Dombek P, Ream W (1997) Functional domains of Agrobacterium tumefaciens single-stranded DNA-binding protein VirE2. J Bacteriol 179:1165–1173

Escudero J, Hohn B (1997) Transfer and integration of T-DNA without cell injury in the host plant. Plant Cell 9:2135–2142

Estruch JJ, Prinsen E, Van Onckelen H, Schell J, Spena A (1991) Viviparous leaves produced by somatic activation of an inactive cytokinin-synthesizing gene. Science 254:1364–1367

Feldmann K (1991) T-DNA insertion mutagenesis in Arabidopsis: mutational spectrum. Plant J 1:71–82

Filichkin SA, Gelvin SB (1993) Formation of a putative relaxation intermediate during T-DNA processing directed by the Agrobacterium tumefaciens VirD1, D2 endonuclease. Mol Microbiol 8:915–926

Finberg KE, Muth TR, Young SP, Maken JB, Heiritter SM, Binns AN, Banta L (1995) Interactions of VirB9, -10, and -11 with the membrane fraction of Agrobacterium tumefaciens: solubility studies provide evidence for tight associations. J Bacteriol 177:4881–4889

Fortin C, Nester EW, Dion P (1992) Growth inhibition and loss of virulence in cultures of Agrobacterium tumefaciens treated with acetosyringone. J Bact 174:5656–5685

Fortin C, Marquis C, Nester EW, Dion P (1993) Dynamic structure of Agrobacterium tumefaciens Ti plasmids. J Bact 175:4790–4799

Fraley RT, Rogers S, Horsch R, Sanders PR, Flick J, Adams SP, Bittner ML, Brand LA, Fink CC, Fry JS, Gallupi GR, Goldberg SB, Hoffmann NL (1983) Expression of bacterial genes in plant cells. Proc Natl Acad Sci USA 80:4803–4807

Fraley RT, Rogers S, Horsch R, Eichholtz D, Flick J, Fink CL, Hoffmann NL, Sanders PR (1985) The SEV system: a new disarmed Ti plasmid vector system for plant transformation. Bio Technology 3:629–635

Fullner KJ (1998) Role of Agrobacterium virB genes in Transfer of T Complexes and RSF1010. J Bacteriol 180:430–434

Fullner KJ, Nester EW (1996) Temperature affects the T-DNA transfer machinery of Agrobacterium tumefaciens. J Bacteriology 178:1498–1504

Fullner KJ, Lara JC, Nester EW (1996) Pilus assembly by Agrobacterium T-DNA transfer genes. Science 273:1107–1109

Ghai J, Das A (1989) The virD operon of Agrobacterium Ti plasmid encodes a DNA relaxing enzyme. Proc Natl Acad Sci 86:3109–3113

Gheysen G, Van Montagu M, Zambryski P (1987) Integration of Agrobacterium tumefaciens transfer DNA (T-DNA) involves rearrangements of target plant DNA sequences. Proc Natl Acad Sci USA 84:6169–6173

Gheysen G, Herman L, Breyne P, Van Montagu M, Depicker A (1990) Cloning and sequence analysis of truncated T-DNA inserts from Nicotiana tabacum. Gene 94:155–163

Gheysen G, Villarroel R, Van Montagu M (1991) Illegitimate recombination in plants: a model for T-DNA integration. Genes Devel 5:287–297

Gielt C, Koukolikova Z, Hohn B (1987) Mobilization of T-DNA from Agrobacterium to plant cells involves a protein that binds single-stranded DNA. Proc Natl Acad Sci USA 84:9006–9010

Grimsley N, Hohn B, Ramos C, Kado C, Rogowsky P (1989) DNA transfer from Agrobacterium to Zea mays or Brassica by agroinfection is dependent on bacterial virulence functions. Mol Gen Genet 217:309–316

Hajdukiewicz P, Svab Z, Maliga P (1994) The small, versatile pPZP family of Agrobacterium binary vectors for plant transformation. Plant Mol Biol 25:989–994

Halfter U, Morris P-C, Willmitzer L (1992) Gene targeting in Arabidopsis thaliana. Mol Gen Genet 231:186–193

Hamilton CM, Frary A, Lewis C, Tanksley SD (1996) Stable transfer of intact high molecular weight DNA into plant chromosomes. Proc Natl Acad Sci USA 93:9975–9979

Hansen G, Chilton MD (1996) "Agrolistic" transformation of plant cells: integration of T-strands generated in planta. Proc Natl Acad Sci USA 93:14978–14983

Hansen G, Larribe M, Vaubert D, Tempé J, Biermann B, Montaya AL, Chilton M-D, Brevet J (1991) *Agrobacterium rhizogenes* pRi8196 T-DNA: mapping and DNA sequence of functions involved in manopine synthesis and hairy root differentiation. Proc Natl Acad Sci USA 88:7763–7767

Hansen G, Tempé J, Brevet J (1992) A T-DNA transfer enhancer sequence in the vicinity of the right border of *Agrobacterium rhizogenes* pRi8196. Plant Mol Biol 20:113–122

Hansen G, Das A, Chilton MD (1994) Constitutive expression of the virulence genes improves the efficiency of plant transformation by Agrobacterium. Proc Natl Acad Sci USA 91:7603–7607

Hansen G, Shillito RD, Chilton MD (1997a) T-strand integration in maize protoplasts after codelivery of a T-DNA substrate and virulence genes. Proc Natl Acad Sci USA 94:11726–11730

Hansen G, Vaubert D, Clérot D, Brevet B (1997b) Wound-inducible and organ-specific expression of ORF13 from *Agrobacterium rhizogenes* 8196 T-DNA in transgenic tobacco plants. Mol Gen Genet 254:337–343

Heath JD, Boulton MI, Raineri DM, Doty SL, Mushegian AR, Charles TC, Davies JW, Nester EW (1997) Discrete regions of the sensor protein VirA determine the strain-specific ability of Agrobacterium to agroinfect maize. MPMI 10:221–227

Herrera-Estrella A, Van Montagu M, Wang K (1990) A bacterial peptide acting as a plant nuclear targeting signal: the amino-terminal portion of Agrobacterium VirD2 protein directs a β- galactosidase fusion protein into tobacco nuclei. Proc Natl Acad Sci USA 87:9534–9537

Hiei Y, Ohta S, Komari T, Kumashiro T (1994) Efficient transformation of rice (*Oryza sativa* L.) mediated by Agrobacterium and sequence analysis of the boundaries of the T-DNA. Plant J 6(2):271–282

Hobbs SLA, Warkentin TD, DeLong CMO (1993) Transgene copy number can be positively or negatively associated with transgene expression. Plant Mol Biol 21:17–26

Hoekema A, Hirsch PR, Hooykaas PJJ, Schilperoort RA (1983) A binary plant vector strategy based on separation of vir and T-region of the *Agrobacterium tumefaciens* Ti-plasmid. Nature 303:179–180

Hood EE, Helmer GL, Fraley RT, Chilton MD (1986) The hypervirulence of *Agrobacterium tumefaciens* A281 is encoded in a region of pTiBo542 outside of T-DNA. J Bacteriol 168:1291–1301

Hooykaas PJJ, Schilperoort RA (1992) Agrobacterium and plant genetic engineering. Plant Mol Biol 19:15–38

Hooykaas PJJ, Hofker M, Den Dulk-Ras H, Schilperoort RA (1984) A comparison of virulence determinants in an octopine Ti plasmid, a nopaline Ti plasmid, and an Ri plasmid by complementation analysis of *Agrobacterium tumefaciens* mutants. Plasmid 11:195–205

Horsch RB, Fry JE, Hoffman NL, Eichholtz D, Rogers SG, Fraley (1985) A simple and general method for transferring genes into plants. Science 227:1229–1231

Howard EA, Zupan JR, Citovsky V, Zambryski P (1992) The VirD2 protein of *Agrobacterium tumefaciens* contains a C-terminal bipartite nuclear localization signal: implications for nuclear uptake of DNA in plant cells. Cell 68:109–118

Iglesias VA, Moscone EA, Papp I, Neuhuber F, Michalowski S, Phelan T, Spiker S, Matzke M, Matzke AJM (1997) Molecular and cytogenetic analyses of stably and unstably expressed transgene loci in tobacco. Plant Cell 9:1251–1264

Ishida Y, Saito H, Ohta S, Hiei Y, Komari T, Kumashiro T (1996) High efficiency transformation of Maize (*Zea mays* L.) mediated by *Agrobacterium tumefaciens*. Nature Biotechnology 14:745–750

Jansen BJ, Gardner RC (1989) Localized transient expression of GUS in leaf discs following cocultivation with Agrobacterium. Plant Mol Biol 14:61–72

Jasper F, Koncz C, Schell J, Steinbiss HH (1994) Agrobacterium T-strand production in vitro: sequence-specific cleavage and 5' protection of single-stranded DNA templates by purified VirD2 protein. Proc Natl Acad Sci 91:694–698

Jefferson RA (1987) Assaying chimeric genes in plants: the GUS gene fusion system. Plant Mol Biol Rep 5:387–405

Jen GC, Chilton MD (1986) The right border region of pTiT37 T-DNA is intrinsically more active than the left border in promoting T-DNA transformation. Proc Natl Acad Sci USA 83:3895–3899

Jin S, Komari T, Gordon MP, Nester EW (1987) Genes responsible for the supervirulence phenotype of *Agrobacterium tumefaciens* A281. J Bacteriol 169:4417–4425

Jin S, Prusti RK, Roitsch T, Ankenbauer, RG Nester EW (1990) Phosphorylation of the VirG protein of *Agrobacterium tumefaciens* by the autophosphorylated VirA protein: essential role in biological activity of VirG. J Bacteriol 172:4945–4950

Jorgensen R, Snyder C, Jones JDG (1987) T-DNA is organized predominantly in inverted repeat structures in plants transformed with *Agrobacterium tumefaciens* C58 derivatives. Mol Gen Genet 207:471–477

Jouanin L, Bouchez D, Drong RF, Tepfer D, Slightom JL (1989) Analysis of TR-DNA plant junctions in the genome of a *Convolvulus arvensis* clone transformed with *Agrobacterium rhizogenes* strain A4 Plant Mol Biol 12:75-85

Kanemoto RH, Powell AT, Akyoshi DE, Reiger DA, Kerstetter RA, Nester EW, Hawes MC, Gordon MP (1989) Nucleotide sequence analysis of the plant-inducible locus pinF from *Agrobacterium tumefaciens*. J Bacteriol 171:2506-2512

Kang HW, Wirawan IGP, Kojima M (1994) Cellular localization and functional analysis of the protein encoded by the chromosomal virulence gene (acvB) of *Agrobacterium tumefaciens*. Biosci Biotech Biochem 58:2024-2032

Kempin SA, Liljegren SJ, Block LM, Rounsley SD, Yanofsky MF, Lam E (1997) Targeted disruption in Arabidopsis. Nature Lond. 389:802-803

Klee H, Horsch R, Rogers S (1987) Agrobacterium-mediated plant transformation and its further applications to plant biology. Ann Rev Plant Physiol 38:467-486

Komari T, Hiei Y, Saito Y, Murai N, Kumashiro T (1996) Vectors carrying two separate T-DNAs for co-transformation of higher plants mediated by *Agrobacterium tumefaciens* and segregation of transformants free from selection markers. Plant J 10:165-174

Koncz C, Schell J (1986) The promoter of T$_L$-DNA gene 5 controls the tissue-specific expression of chimaeric genes carried by a novel type of Agrobacterium binary vector. Mol Gen Genet 204:383-396

Koncz C, Martini N, Meyerhofer R, Koncz-Kalman Z, Körber H, Redei GP, Schell J (1989) High-frequency T-DNA-mediated gene tagging in plants. Proc Natl Acad Sci USA 86:8467-8471

Koncz C, Nemeth K, Redei GP, Schell J (1994) Homology recognition during T-DNA integration into the plant genome. Homologous Recombination and Gene Silencing in Plants. Ed. J Paszkowski. Netherlands, pp 167-189

Kononov ME, Bassuner B, Gelvin SB (1997) Integration of T-DNA binary vector backbone sequences into the tobacco genome-Evidence for multiple complex patterns of integration. Plant J 11:945-957

Lai, E, Kado CI (1998) Processed VirB2 is the major subunit of the promiscuous pilus of *Agrobacterium tumefaciens*. J Bacteriol 180:2711-2717

Lam ST, Torkewitz NR, Nautiyal CS, Dion P (1997) Micro-organisms with mannopine catabolizing activity Patent number 5610044 USA Date of patent Mar 11, 1997

Lee KY, Lund P, Lowe K, Dunsmuir P (1990) Homologous recombination in plant cells after Agro-bacterium-mediated transformation. Plant Cell 2:415-425

Leroux B, Yanofsky MF, Winans SC, Ward JE, Ziegler SF, Nester EW (1987) Characterization of the virA Locus of *Agrobacterium tumefaciens*: a transcriptional regulator and host range determinant. EMBO J 6:849-856

Lessl M, Lanka E (1994) Common mechanisms in bacterial conjugation and Ti-mediated transfer to plant cells. Cell 77:321-324

Lippincott BB, Lippincott JA (1969) Bacterial attachment to a specific wound site as an essential stage in tumor initiation by *Agrobacterium tumefaciens*. J Bact 97:620-628

Martienssen RA (1998) Functional genomics: probing plant gene function and expression with trans-posons. Proc Natl Acad Sci USA 95:2021-2026

Matsumoto S, Ito Y, Hodoi T, Takahashi Y, Machida Y (1990) Integration of Agrobacterium T-DNA into a tobacco chromosome: possible involvement of DNA homology between T-DNA and plant DNA. Mol Gen Genet 224:309-316

Matthysse AG (1986) Initial interactions of *Agrobacterium tumefaciens* with plant host cells. CRC Crit Rev Microbiol 13:281-307

Matthysse AG, White S, Lightfoot R (1995a) Genes required for cellulose synthesis in *Agrobacterium tumefaciens*. J Bacteriol 177:1069-1075

Matthysse AG, Thomas DL, White AR (1995b) Mechanism of cellulose synthesis in *Agrobacterium tumefaciens*. J Bact 177:1076-1081

Matthysse AG, Yarnall HA, Young N (1996) Requirements for gene with homology to ABC transport systems for attachment and virulence of *Agrobacterium tumefaciens*. J Bacteriol 178:5302-5308

Mayerhofer R, Koncz-Kalman Z, Nawrath C, Bakkeren G, Crameri A, Angelis K, Redei GP, Schell J, Hohn B, Koncz C (1991) T-DNA integration: a mode of illegitimate recombination. EMBO J 10:697-704

McKnight TD, Lillis MT, Simpson RB (1987) Segregation of genes transferred to one plant cell from two separate Agrobacterium strains. Plant Mol Biol 8:439-445

Melchers LS, Maroney MJ, den Dulk-Ras A, Thompson DV, van Vuuren HAJ, Schilperoort RA, Hooykaas PJJ (1990) Octopine and nopaline strains of *Agrobacterium tumefaciens* differ in virulence; molecular characterization of the virF locus. Plant Mol Biol 14:249-259

Miao A-H, Lam E (1995) Targeted disruption of the TGA3 locus in Arabidopsis thaliana. Plant J 7(2):359–365

Mlynarova L, Keiser LCP, Stiekema W, Nap J-P (1996) Approaching the lower limits of transgene variability. Plant Cell 8:1589–1599

Mysore KS, Bassuner B, Deng X-B, Darbinian NS, Motchoulski A, Ream W, Gelvin SB (1998) Role of the *Agrobacterium tumefaciens* VirD2 protein in T-DNA transfer and integration. Mol Plant Microbe Interact 11:668–683

Nam J, Matthysse A, Gelvin S (1997) Differences in susceptibility of Arabidopsis ecotypes to crown gall disease may result from a deficiency in T-DNA integration. The Plant Cell 9:317–333

Narasimhulu SB, Deng X-b, Sarria R, Gelvin SB (1996) Early transcription of Agrobacterium T-DNA genes in tobacco and maize. Plant Cell 8:873–886

Ni M, Cui D, Einstein J, Narasimhulu S, Vergera CE, Gelvin SB (1995) Strength and tissue specificity of chimeric promoters derived from the octopine and mannopine synthase genes. Plant J 7:661–676

Offringa R, de Groot MJA, Haagman HJ, Does MP, van den Elzen PJM, Hooykaas PJJ (1990) Extrachromosomal homologous recombination and gene targeting in plant cells after Agrobacterium-mediated transformation. EMBO J 9:3077–3084

Offringa R, Franke-van Dijk MEI, de Groot MJA, van den Elzen PJM, Hooykaas PJJ (1993) Non-reciprocal homologous recombination between Agrobacterium T-DNA and a plant chromosomal locus. Proc Natl Acad Sci USA. 90:7346–7350

Oger P, Petit A, Dessaux Y (1997) Genetically engineered plants producing opines alter their biological environment. Nature Biotechnology 15:369–372

Ohta S, Mita S, Hattori T, Nakamura K (1990) Contruction and expression in tobacco of a ββ-glucuronidase (GUS) reporter gene containing an intron within the coding sequence. Plant Cell Physiol 31:805–813

Otten L, De Greve H, Leemans J, Hain R, Hooykaas P, Schell J (1984) Restoration of virulence of Vir region mutants of *Agrobacterium tumefaciens* strain B6S3 by coinfection with normal and mutant Agrobacterium strains. Mol Gen Genet 195:159–163

Otten L, Piotrowiack G, Hooykaas P, Dubois M, Szegedi E, Schell J (1985) Identification of an *Agrobacterium tumefaciens* pTiB6S3 vir region fragment that enhances the virulence of pTiC58. Mol Gen Genet 199:189–193

Otten L, Canaday J, Gérard JC, Fournier P, Crouzet P, Paulus F (1992) Evolution of Agrobacteria and their Ti plasmids-a review. Mol Plant Microbe Interact 5:279–287

Pansegrau W, Schoumacher F, Hohn B, Lanka E (1993) Site-specific cleavage and joining of single-stranded DNA by VirD2 protein of *Agrobacterium tumefaciens* Ti plasmids: Analogy to bacterial conjugation. Proc Natl Acad Sci 90:11538–11542

Paszkowski J, Baur M, Bogucki A, Potrykus I (1988) Gene targeting in plants. EMBO J 13:4021–4026

Pazour GP, Ta VN, Das A (1992) Constitutive mutations of *Agrobacterium tumefaciens* transcriptional activator virG. J Bact 174:4169–4174

Peralta GE, Helmiss R, Ream W (1986) "overdrive", a T-DNA transmission enhancer on the *Agrobacterium tumefaciens* tumor-inducing plasmid. EMBO J 5:1137–1142

Perl A, Lotan O, Abu-Abied M, Holland D (1996) Establishment of an Agrobacterium-mediated transformation system for grape (Vitis vinifera L.): the role of antioxidants during grape-Agrobacterium interactions. Nature Biotech 14:624–628

Petit A, Tempé J, Kerr A, Holsters M, Van Montagu M, Schell J (1978) Substrate induction of conjugative activity of Agrobacterium tumefaciens Ti plasmids. Nature 271:570–571

Porter SG, Yanofsky MF, Nester EW (1987) Molecular characterization of the virD operon from *Agrobacterium tumefaciens*. Nucleic Acids Res 15:7503–7515

Puchta H, Dujon B, Hohn B (1996) Two different but related mechanisms are used in plants for the repair of genomic double-strand breaks by homologous recombination. Proc Natl Acad Sci USA 93:5055–5060

Ramanathan V, Veluthambi K (1995) Transfer of non-T-DNA portions of the *Agrobacterium tumefaciens* Ti plasmid pTiA6 from the left terminus of TL-DNA. Plant Mol Biol 28:1149–1154

Ream W (1989) *Agrobacterium tumefaciens* and interkingdom genetic exchange. Ann Rev Phytopathol 1989:583–618

Regenburg-Tu AJG, Hooykaas PJJ (1993) Transgenic N glauca plants expressing bacterial virulence gene virF are converted into hosts for nopaline strains. Nature (Lond)363:69–71

Reuhs B, Kim JS, Matthysse AG (1997) Attachment of *Agrobacterium tumefaciens* to carrot cells and Arabidopsis wound sites is correlated with the presence of a cell-associated, acidic polysaccharide. J Bacteriol 179:5372–5379

Risseeuw E, Offringa R, Franke-van Dijk M, Hooykaas PJJ (1995) Targeted recombination in plants using Agrobacterium coincides with additional rearrangements at the target locus. Plant J 7: 109–119

Risseeuw E, Franke-van Dijk M, Hooykaas PJJ (1997) Gene targeting and instability of Agrobacterium T-DNA loci in the plant genome. Plant J 11:717–728

Risuleo G, Battistoni P, Costantino P (1982) Regions of homology between tumorigenic plasmids from Agrobacterium rhizogenes and Agrobacterium tumefaciens. Plasmid 7:45–51

Rossi L, Escudero J, Hohn B, Tinland B (1993) Efficient and sensitive assay for T-DNA-dependent transient gene expression. Plant Mol Biol Reporter 11:220–229

Rossi L, Hohn B, Tinland B (1993) The VirD2 protein of Agrobacterium tumefaciens carries nuclear localization signals important for transfer of T-DNA to plants. Mol Gen Genet 239:345–353

Rossi L, Hohn B, Tinland B (1996) Integration of complete DNA units is dependent on the activity of virulence E2 protein of Agrobacterium tumefaciens. Proc Natl Acac Sci USA 93:126–130

Rossi L, Tinland B, Hohn B (1998) Role of virulence proteins of Agrobacterium in the plant, in press in The Rhizobiaceae, eds. Spaink H, Hooykaas P, Kondorosi A, Kluwer. Dordrecht, The Netherlands, pp 303–320

Russel SH, Hoopes JL, Odell JT (1992) Directed excision of a transgene from the plant genome. Mol Gen Genet 234:49–59

Sahi S, Chilton MD, Chilton W S (1990) Corn metabolites affect growth and virulence of Agrobacterium tumefaciens Proc Natl Acad Sci USA 87:3879–3883

Savka MA, Farrand S (1997) Modification of rhizobacterial populations by engineering bacterium utilization of a novel plant-produced resource. Nature Biotech 15:363–368

Schaefer DG, Zryd JP (1997) Efficient gene targeting in the moss Physcomitrella patens. Plant J 11: 1195–1206

Scheiffele P, Pansegrau W, Lanka E (1995) Initiation of Agrobacterium tumefaciens T-DNA processing. J Biol Chemistry 270:1269–1276

Sederoff R, Stomp A-M, Chilton WS, Moore LW (1986) Gene transfer into loblolly pine by Agrobacterium tumefaciens. Bio/Tech 4:647–649

Shaw CH (1984) The right hand copy of the nopaline Ti plasmid 25 bp repeat is required for tumor formation. Nucleic Acids Res 12:6031–6041

Shen W H, Escudero J, Schlappi M, Ramos C, Hohn B, Koukolilova-Nicola Z (1993) T-DNA transfer to maize cells: Histochemical investigation of β-glucuronidase activity in maize tissues. Proc Natl Acad Sci USA 90:1488–1492

Sheng J, Citovsky V (1996) Agrobacterium-plant cell DNA transport: have virulence proteins, will travel. The Plant Cell 8:1699–1710

Shurvinton CE, Hodges L, Ream W (1992) A nuclear localization signal and the C-terminal omega sequence in the Agrobacterium tumefaciens VirD2 endonuclease are important for tumor formation. Proc Natl Acad Sci USA 89:11837–11841

Slightom JL, Durand-Tardif M, Jouanin L, Tepfer D (1986) Nucleotide sequence analysis of TL-DNA Agrobacterium rhizogenes type plasmid. Identification of open reading frames. J Biol Chem 261:108–121

Smigocki AC, Owens LD (1988) Cytokinin gene fused with a strong promoter enhances shoot organogenesis and zeatin levels in transformed plant cells. Proc Natl Acad Sci USA 85:5131–5135

Smit G, Swart S, Lugtenberg BJJ, Kijne JW (1992) Molecular mechanisms of attachment of Rhizobium bacteria to plant roots. Mol Microbiol 6:2897–2903

Sonti RV, Chiurazzi M, Wong D, Davies CS, Harlow GR, Mount DW, Signer ER (1995) Arabidopsis mutants deficient in T-DNA integration. Proc Natl Acad Sci USA 92:11786–11790

Stachel SE, Messens E, Van Montagu M, Zambryski P (1985) Identification of the signal molecules produced by wounded plant cells that activate T-DNA transfer in Agrobacterium tumefaciens. Nature 318:624–629

Stachel SE, Nester EW (1986) The genetic and transcriptional organization of the vir region of the A6 Ti plasmid of Agrobacterium tumefaciens. EMBO J 5:1445–1454

Stachel SE, Timmerman B, Zambryski P (1986) Generation of single-stranded T-DNA molecules during the initial stages of T-DNA transfer for Agrobacterium tumefaciens to plant cells. Nature 322:706–712

Stachel SE, Timmerman B, Zambryski P (1987) Activation of Agrobacterium tumefaciens vir gene expression generates multiple single-stranded T-strand molecules from the pTiA6 T-region: requirements for 5'virD gene products. EMBO J 6:857–863

Stomp A-M, Loopstra C, Chilton WS, Sederoff RR, and Moore LW (1990) Extended host range of Agrobacterium tumefaciens in the genus Pinus. Plant Physiol 92:1226–1232

Sundaresan V, Springer P, Volpe T, Haward S, Jones JDG, Dean C, Ma H, Martienssen R (1995) Patterns of gene action in plant development revealed by enhancer trap and gene trap transposable elements. Genes Develop 9:1797–1810

Sundberg C, Meek L, Carrol K, Das A, Ream W (1996) VirE1 protein mediates export of single-stranded DNA binding protein VirE2 from *Agrobacterium tumefaciens* into plant cells. J Bacteriol 178:1207–1212

Swart S, Lugtenberg BJJ, Smit G and Kijne JW (1994a) Rhicadhesin-mediated attachment and virulence of an *Agrobacterium tumefaciens* chvB mutant can be restored by growth in a highly osmotic medium. J Bacteriol 176:3816–3819

Swart S, Logman, T JJ, Smit G, Lugtenberg BJJ, Kijne JW (1994b) Purification and partial characterization of a glycoprotein from pea (Pisum sativum) with receptor activity for rhicadhesin, an attachment protein of Rhizobiaceae. Plant Mol Biol 24:171–183

Szegedi E, Korbury J, Koleda I (1984) Crown gall resistance in East-Asian Vitis species and their V. vinifera hybrids. Vitis 23:21–26

Thomashow MF, Karlinsey JE, Marks JR, Hurlbert RE (1987) Identification of a new virulence locus in *Agrobacterium tumefaciens* that affects polysaccharide composition and plant cell attachment. J Bact 169:3209–3216

Thorstenson YR, Kuldau GA, Zambryski PC (1993) Subcellular localization of seven VirB proteins of *Agrobacterium tumefaciens*: implications for the formation of a T-DNA transport structure. J Bacteriol 175:5233–5241

Thykjaer T, Finemann J, Schauser L, Christensen L, Poulsen C, Stougaard J (1997) Gene targeting approaches using positive-negative selection and large flanking regions. Plant Mol Biol 35:523–530

Tinland B, Hohn B, Puchta H (1994) *Agrobacterium tumefaciens* transfers single-stranded transferred DNA (T-DNA) into the plant cell nucleus. Proc Natl Acad Sci USA 91:8000–8004

Tinland B, Koukolíková-Nicola Z, Hall MN, Hohn B (1992) The T-DNA-linked VirD2 protein contains two distinct functional nuclear localization signals. Proc Natl Acad Sci USA 89:7442–7446

Tinland B, Schoumacher F, Gloeckler V, Bravo-Angel AM, Hohn B (1995) The *Agrobacterium tumefaciens* virulence D2 protein is responsible for precise integration of T-DNA into the plant genome. EMBO J 14:3585–3595

Torisky RS, Kovacs L, Avdiushko A, Newman JD, Hunt AG, Collins GB (1997) Development of a binary vector system for plant transformation based on the supervirulent *Agrobacterium tumefaciens* strain Chry5. Plant Cell Reports 17:102–108

Toro N, Data A, Yanofsky M, Nester EW (1988) Role of the overdrive sequence in T-DNA border cleavage in Agrobacterium. Proc Natl Acad Sci USA 85:8558–8562

Toro N, Data A, Carmi OA, Young C, Prusti RK, Nester EW (1989) The *Agrobacterium tumefaciens* virC 1 gene product binds to overdrive, a T-DNA transfer enhancer. J Bacteriol 171:6845–6849

Trick HN, Finer JJ (1998) Sonication-assisted Agrobacterium-mediated transformation of soybean [Glycine max (L.) Merill] embryogenic suspension culture tissue. Plant Cell Rep 17:482–488

Uknes S, Dinder S, Friedrich L, Negrotto D, Williams S, Thompson-Taylor H, Potter S, Ward E, Ryals J (1993) Regulation of pathogenesis-related protein-1a gene expression in tobacco. Plant Cell 5:159–169

Vancanneyt G, Schmidt R, O'Connor-Sanchez A, Willmitzer L, Rocha-Sosa M (1990) Construction of an intron-containing marker gene: splicing of the intron in transgenic plants and its use in monitoring early events in Agrobacterium-mediated plant transformation. Mol Gen Genet 220:245–250

Van de Broek A, Vanderleyden J (1995) The role of bacterial motility, chemotaxis, and attachment in bacterial-plant interactions. MPMI 8:800–810

Van Haaren MJJ, Sedee NJA, Schilperoort RA, Hooykaas PJJ (1987) Overdrive is a T-region transfer enhancer which stimulates T-strand production in A. tumefaciens. Nucleic Acids Res 15:8983–8997

Van Larebeke N, Engler G, Holsters M, Van den Elsacker S, Zaenen I, Schilperoort RA, Schell J (1974) Large plasmid in *Agrobacterium tumefaciens* essential for crown gall inducing ability. Nature (Lond) 252:169–170

Vergunst AC, Jansen LET, Hooykaas PJJ (1998) Site-specific integration of Agrobacterium T-DNA in Arabidopsis thaliana mediated by Cre recombinase. Nucleic Acids Res 26:2729–2734

Wagner V, Matthysse AG (1992) Involvment of a vitronectin-like protein in attachment of *Agrobacterium* tumefaciens to carrot suspension culture cells. J Bacteriol 174:5999–6003

Wang K, Genetello C, Van Montagu M, Zambryski P (1987a) Sequence context of the T-DNA border repeat element determines its relative activity during T-DNA transfer to plant cells. Mol Gen Genet 210:338–346

Wang K, Stachel SE, Timmerman B, Van Montagu M, Zambryski P (1987b) Site-specific nick occurs within the 25-bp transfer promoting border sequence following induction of vir gene expression in *Agrobacterium tumefaciens*. Science 235:587–591

Wang K, Herrera-Estrella A, Van Montagu M (1990) Overexpression of virD1 and virD2 genes in *Agrobacterium tumefaciens* enhances T-complex formation and plant transformation. J Bacteriol 172:4432–4440

Watson B, Currier TC, Gordon MP, Chilton M-D, Nester EW (1975) Plasmid required for virulence of *Agrobacterium tumefaciens*. J Bacteriol 123:255–264

Wenck A, Czako M, Kanevski I, Marton L (1997) Frequent co-linear long transfer of DNA inclusive of the whole binary vector during Agrobacterium-mediated transformation. Plant Mol Biol 34:915–922

White FF, Nester EW (1980) Hairy root: plasmid encodes virulence traits in *Agrobacterium rhizogenes*. J Bacteriol 141:1134–1141

White FF, Taylor BH, Huffman GA, Gordon MP, Nester EW (1985) Molecular and genetic analysis of the transferred DNA regions of the root inducing plasmid of *Agrobacterium rhizogenes*. J Bacteriol 164:33–44

Winans SC, Allenza P, Stachel SE, McBride KE, Nester EW (1987) Characterization of the virE operon of the Agrobacterium Ti plasmid pTiA6. Nucleic Acids Res 15:825–837

Wirawan IGP, Kojima M (1996) Chromosomal virulence gene (acvB) product of *Agrobacterium tumefaciens* that binds to a T-strand to mediate its transfer to host plant cells. Biosci Biotech Biochem 60:44–49

Wirawan IGP, Kang HW, Kojima M (1993) Isolation and characterization of a new chromosomal virulence gene of *Agrobacterium tumefaciens*. J Bacteriol 175:3208–3212

Yadav NS, Vanderleyden J, Bennett DR, Barnes WM, Chilton M-D (1982) Short direct repeats flank the T-DNA on a nopaline Ti plasmid. Proc Natl Acad Sci USA 79:6322–6326

Yanofsky MF, Nester EW (1986) Molecular characterization of a host-range-determining locus from *Agrobacterium tumefaciens*. J Bacteriol 168:237–243

Yanofsky M, Lowe B, Montaya A, Rubin R, Krul W, Gordon M, Nester EW (1985) Molecular and genetic analysis of factors controlling host range in *Agrobacterium tumefaciens*. Mol Gen Genet 201:237–246

Yanofsky MF, Porter SG, Young C, Albright LM, Gordon MP (1986) The virD operon of *Agrobacterium tumefaciens* encodes a site specific endonuclease. Cell 47:471–477

Yoder JI, Goldsbrough AP (1994) Transformation systems for generating marker-free transgenic plants. Bio/Technology 12:263–268

Yusibov VM, Steck TR, Gupta V, Gelvin SB (1994) Association of single-stranded DNA from *Agrobacterium tumefaciens* with tobacco cells. Proc Natl Acad Sci 91:2994–2998

Zaenen I, Van Larebeke N, Teuchy H, Van Montagu M, Schell J (1974) Supercoiled circular DNA in crown-gall-inducing Agrobacterium strains. J Mol Biol 86:109–127

Zambryski P (1988) Basic processes underlying Agrobacterium-mediated DNA transfer to plant cells. Annu Rev Genet 22:1–30

Zambryski P (1992) Chronicles from the Agrobacterium-plant cell DNA transfer story. Ann Rev Plant Physiol Plant Mol Biol 43:465–490

Zambryski P, Joos H, Genetello C, Leemans J, Van Montagu M, Schell J (1983) Ti plasmid vector for the introduction of DNA into plant cells without alteration of their normal alteration capacity. EMBO J 2:2143–2150

Zorreguieta A, Ugalde RA (1986) Formation in Rhizobium and Agrobacterium spp of a 235-kilodalton protein intermediate in β-d-(1,2)glucan synthesis. J Bacteriol 167:947–941

Zupan JR, Zambryski PC (1995) Transfer of T-DNA from Agrobacterium to the plant cell. Plant Physiol 107:1041–1047

Zupan JR, Citovsky V, Zambryski P (1996) Agrobacterium VirE2 protein mediates nuclear uptake of single-stranded DNA in plant cells. Proc Natl Acad Sci USA 93:2392–2397

Particle Bombardment Mediated Transformation

J.J. FINER[1], K.R. FINER[2], and T. PONAPPA[1]

[1] Department of Horticulture and Crop Science, Plant Molecular Biology and Biotechnology Program, Ohio Agricultural Research and Development Center, The Ohio State University, Wooster, OH 44691, USA

[2] Department of Biological Sciences, Kent State University, Stark Campus, Canton, OH 44720, USA

1 Introduction

Since the production of the first genetically engineered plants in the middle 1980s (HORSCH et al. 1985), there has been an explosion of interest in the production of transgenic plants for both basic and applied work. Large efforts have been placed in the development and refinement of methods for transgenic plant production and also in the isolation and characterization of useful genes for introduction into plants. Through gene transfer, plants have already been produced that contain genes for disease and insect resistance, and modified fruit and grain quality. Many of the products of plant biotechnology would not exist today without particle bombardment gene transfer methods. Using Biolistics, transformation of many plants and tissues that had been recalcitrant to manipulation, has become much more routine. Particle bombardment or Biolistics has been used to transform a large number of different plant species (reviewed by CHRISTOU 1994), as well as some animals, fungi, and bacteria.

2 History

A new method using high-velocity microprojectiles to deliver exogenous genetic material into plant tissue was developed and first described by John Sanford in collaboration with Ted Klein, Ed Wolf, and Nelson Allen (SANFORD et al. 1987). In that first study, penetration of particles was easily visualized using onion tissue, where a monolayer of large epidermal cells was used as the target tissue. By observing cytoplasmic streaming in these cells, this tissue could be easily evaluated for viability following bombardment. Onion epidermal cells, transformed with the chloramphenicol acetyl transferase (CAT) gene (KLEIN et al. 1987), were the first transiently transformed cells produced using this method. CAT activity was detected in the tissue 3 days after delivery of particles using an early gunpowder discharge device.

The first stably transformed plants obtained using biolistics were reported by CHRISTOU et al. (1988) and KLEIN et al. (1988) using soybean and tobacco respectively. CHRISTOU et al. (1988, 1989) bombarded soybean shoot meristems and recovered chimeric plants which subsequently transmitted the introduced gene to progeny. KLEIN et al. (1988) targeted tobacco leaf tissue which formed callus under kanamycin selection. Plants were eventually recovered from the tobacco callus using standard regeneration protocols. Since these initial reports of successful transformation, a broad range of species has been transformed including bacteria (SMITH et al. 1992), fungi (TOFFALETTI et al. 1993) and even animals (JOHNSTON et al. 1991). In addition, organelles such as chloroplasts (BOYNTON et al. 1988) and mitochondria (JOHNSTON et al. 1988) have also been transformed using Biolistics.

This method of particle acceleration, which has been described as the particle gun, the gene gun, the bioblaster, and the microprojectile bombardment method was initially christened the "biolistic" method by its inventor (SANFORD 1988). Biolistics, a combination of "biological" and "ballistics", describes a technique which utilizes instrumentation to accelerate DNA coated microprojectiles into cells, past the cell wall and cell membrane. The microprojectile is small enough (0.5–5 μm) to enter the plant cell without too much damage, yet large enough to have the mass to penetrate the cell wall and carry an appropriate amount of DNA on its surface into the interior of the plant cell.

Because this is a direct gene transfer method, which does not rely on a biological vector, Biolistics has been used successfully to transform cells that were once impossible to transform by other means. With this technique, the microprojectile is shot through the physical barriers of the cell, and as such, there are no biological limitations to DNA delivery. Transformation via the Biolistics method is simple and safe and DNA delivery appears to be genotype and tissue independent.

The utility of the Biolistic method is broad and the technology can be used for basic studies on transformation (JOHNSTON et al. 1988), modifications of gene expression (NAPOLI et al. 1990), understanding essential components of genes (MONTGOMERY et al. 1993) and the production of transgenic plants of economic value (FINER and MCMULLEN 1990; GORDON-KAMM et al. 1990; SOMERS et al. 1992; CHRISTOU et al. 1991; BOWER and BIRCH 1992; WEEKS et al. 1993).

3 Particle Gun Design

There are a number of different particle gun designs that are in use in various laboratories. The basis of all of these designs is to coat DNA onto small dense particles and accelerate the particles towards a target tissue. The particles usually consist of either gold or tungsten. Gold particles are chemically inert, produce no cytotoxic effects and are more uniform in size than tungsten particles. However, the cost of gold particles may limit their accessibility and the more affordable tungsten particles are sufficient for most studies. Tungsten particles are somewhat phytotoxic (RUSSELL et al. 1992a) and tend to be more variable in size than gold particles. Ideally the particles used for bombardment should have good initial affinity for the DNA, yet freely release it once in the cytoplasm or nucleus of the target cell.

To prepare DNA-coated microprojectiles, washed gold or tungsten particles are mixed with plasmid DNA. The DNA is bound on the particles using either ethanol or $CaCl_2$ precipitation methods. Spermidine may be added to the mixture, possibly protecting the DNA from degradation and/or altering its conformation. After precipitation, the particles may be washed, resuspended and either dried or stored on ice as an aqueous suspension until use.

Many of the particle guns utilize a macrocarrier, which supports or carries the particles and is accelerated along with the particles towards the target. The macrocarrier is usually retained by a stopping plate or screen before it collides with the target, whereas the particles continue along their course. In most cases, the particles are accelerated under partial vacuum in a vacuum chamber to reduce air drag. Particle penetration is controlled by modifying the intensity of the explosive burst, by changing the distance that the particles must travel to reach the target tissue or by using different sized particles. Most laboratories use the commercially available devices which can be purchased from BioRad Laboratories (Hercules, CA, USA). Depending on accessibility, financial limitations and specific use, it may be appropriate to consider other devices.

3.1 The Original Particle Gun

The original gene gun described in the literature used DNA-coated tungsten particles in an aqueous slurry loaded on the end of a small plastic bullet, which served as the macrocarrier (SANFORD et al. 1987). The plastic macrocarrier was placed into a 0.22 caliber barrel in front of a gun powder cartridge. Once the cartridge was fired, the plastic bullet was propelled towards a solid stopping plate containing a small opening in the center. The microprojectiles continued through the hole towards the target tissue while the macrocarrier was retained. Both the stopping plate and the macrocarrier were discarded afterwards. The initial blast from the gunpowder cartridge necessitated frequent cleaning of the chamber and the target tissue was frequently damaged by the force of the blast. Although large amounts of variation in results were common with this device, this first unit was used successfully by a number of different laboratories.

3.2 Helium-Modified Bombardment Device

The first important modification of the original device included the substitution of a helium blast for the gunpowder discharge (SANFORD et al. 1991). This device, licensed by duPont and marketed by BioRad (Hercules, CA) as the "PDS-1000/He", uses high-pressure helium as the source of particle propulsion. In the PDS-1000/He device, helium pressure continually builds in a reservoir and is released using "rupture discs" which are manufactured to rupture at predetermined pressures. The release of helium produces a shockwave which travels to a second disc or macrocarrier that holds the DNA-coated particles. The macrocarrier carries the DNA coated particles a short distance into a stopping screen which retains the macrocarrier. The microprojectiles continue traveling, ultimately penetrating the target tissue which is held in a partial vacuum. The force of the original shockwave and consequent velocity of the particles can be controlled by using various rupture discs that are made to rupture at different pressures. The PDS-1000/He is the most commonly used device and has shown good control of particle acceleration using the various rupture discs.

3.3 Accel Particle Gun

The Accel gene gun uses a high voltage electrical discharge to vaporize a water droplet which produces a controlled shock wave (MCCABE and CHRISTOU 1993). The device is constructed so that the initial shock wave is reflected to produce secondary shock waves, which in turn accelerate a mylar sheet coated with particles. The mylar sheet is accelerated towards a retaining screen which stops the carrier and allows the particles to continue onto the target tissue. Particle speed, and the resulting depth of particle penetration, may be accurately modified by adjusting the intensity of the voltage passing through the water droplet. The Accel device may be the most efficient particle bombardment device in current use but it is not commercially available and has not received widespread use.

3.4 Particle Inflow Gun

The cost of the above devices (due to manufacturing expenses, patenting constraints and licensing) have placed this technology out of reach for many laboratories. This created a need to develop more affordable, simple, yet safe and effective devices such as the Particle Inflow Gun or PIG (FINER et al. 1992). With this device, a solenoid controlled by a timer relay is used to generate a burst of low-pressure helium. The particles, supported by a screen in a reusable syringe filter unit, are accelerated directly in a stream of helium without the need for a macrocarrier. The target tissue is held in a chamber under a partial vacuum. To minimize the localized impact of the particles and to aid in dispersal, a nylon mesh baffle may be placed over the tissue. Advantages of this design include affordability (no purchase of macrocarriers or rupture membranes), less cleanup and minimal down time between shots.

3.5 Microtargeting Device

Another particle bombardment device that does not utilize a macrocarrier is the microtargeting device (SAUTTER et al. 1991). This device, which accelerates small amounts of a DNA/particle mixture in a focused stream of high-pressure nitrogen, was originally developed to allow precise delivery of particles to the shoot meristem. DNA is not precipitated on the gold particles but is used as a DNA/particle mixture. If the shoot meristem could be accurately targeted, outgrowth of the resultant chimeral or transgenic shoot could rapidly give transgenic progeny.

Transient expression studies using GUS with the shoot meristem as the target indicated that the microtargeting apparatus had great potential for apical meristem transformation (SAUTTER et al. 1991). Particle delivery was precisely controlled and transient expression was limited to the meristematic area. Unfortunately, recovery of transgenic shoots and progeny was never obtained following microtargeting of the shoot meristem. Transformed plants were obtained but not from direct meristem transformation.

If the problems of controlling particle penetration could be overcome and large amounts of meristems could be prepared and individually bombarded, it is conceivable that this device could be effective. However, this is unlikely as only a few microtargeting devices were built and this approach is receiving little attention at present.

3.6 Helios Gene Gun

While the above particle guns targeted tissue held in a vacuum, these apparatuses are not very portable and could not be easily used to transform target tissue that was sensitive to those conditions created by an evacuated chamber. The need for a gene gun which could be hand-held and used without vacuum proved the impetus for an additional gun design.

The Helios Gene Gun (BioRad, Hercules, CA) is a hand-held device which uses low-pressure helium to accelerate DNA coated particles from the interior of a small plastic cartridge towards the target tissue. A spacer at the tip of the gun maintains optimal target distance and minimizes cell damage by venting the helium gas away from the tissue. The Helios Gene Gun can accommodate up to 12 loaded cartridges at once thus allowing for multiple firing of the device before reloading is necessary. The greatest utility of this device seems to be *in situ* transformation of animal cells and tissues (SUNDARAM et al. 1996).

4 Comparison to *Agrobacterium*-Mediated Transformation

The main advantage of particle bombardment over *Agrobacterium* is absence of the biological incompatibilities found when using this biological vector. In the plant kingdom, particle bombardment has shown good utility for transformation of conifers, dicots and monocots. However, particle bombardment may not be the method of choice for all gene transfer work in plants as Biolistics does have some drawbacks relating to cost, ease of use, accessibility and end product utility.

Agrobacterium tumefaciens is a gram-negative, soil-borne pathogenic bacterium that has the unique and natural ability to transfer part of its DNA to the cells of a host plant (CHILTON et al. 1977). Unlike Biolistics, transformation with *Agrobacterium tumefaciens* relies on a host-parasite interaction in order to be successful. *Agrobacterium* appears to have a limited host range with a preference for some dicots while other dicots and most monocots and conifers are, in general, more recalcitrant to *Agrobacterium*-mediated transformation. Although these limitations have been tremendously reduced with our increased knowledge of the biological vector (HANSEN et al. 1994; HIEI et al. 1994; ISHIDA et al. 1996; STACHEL et al. 1985), host range limitations are still present and do present a barrier to use of *Agrobacterium* for many plants. In addition, infection of the target tissue requires wounding of that tissue and elimination of *Agrobacterium* from inoculated tissues may prove problematic.

The main advantages of *Agrobacterium* are ease of inoculation and cost. Delivery of *Agrobacterium* is simple and straightforward, and there are no equipment costs involved. Another advantage of *Agrobacterium* lies in the characterization of transgenics. Transgenic plants obtained via *Agrobacterium* generally contained more predictable introduced DNAs (TINLAND and HOHN 1995) while particle bombardment, as well as other direct DNA uptake methods give rise to more random and uncontrolled DNA integration events (HADI et al. 1996).

5 Factors Affecting Particle Bombardment-Mediated Transformation

Particle bombardment is a physical process of DNA introduction. After the DNA-coated particles are introduced, there are numerous factors that influence gene expression and the subsequent recovery of transgenic plants. These factors are logically related to the ability of the target tissue to process the introduced DNA, resulting in the expression and eventual integration of the foreign DNA. For successful transformation via particle bombardment, the plasmid DNA should be carefully designed and selected, while special attention also needs to be placed on the nature, state and receptivity of the target tissues.

5.1 Components of the Introduced Plasmid DNA

The introduced plasmid DNA(s) typically contains the necessary plasmid backbone (which allows growth and selection in the bacterial host), plant marker genes and the gene(s) of interest.

The "plant marker genes" can be selectable markers (which allow growth of transformed plant cells in the presence of a selective agent) and/or scorable markers (reporter genes). The genes themselves are composed of the promoter, coding region and terminator. Introns, enhancers and attachment regions are defined regions of additional DNA that can be used to enhance, modulate or stabilize gene expression.

5.1.1 Selectable Marker Genes

Choosing the proper selectable marker gene is critical for the successful recovery of stably transformed tissues. The selectable marker simply allows the transformed cells to proliferate in the presence of a selective agent while nontransformed cells either do not grow or multiply at a much reduced rate. Selectable marker genes are usually either antibiotic or herbicide resistance genes. The resistance genes can act either by modifying the selective agent itself to inactivate it or by producing an insensitive form of the normal target of the selective agent.

The most commonly used selectable marker gene is NPTII which imparts resistance to the antibiotic kanamycin (SEKI et al. 1991; RUSSELL et al. 1992b). The NPTII gene product is neomycin phosphotransferase, which inactivates the antibiotic by phosphorylation. Although kanamycin is used most often with NPTII, analogues such as geneticin (G418) and paramomycin are also used with this gene. For some tissues, use of kanamycin with the NPTII gene is adequate while geneticin or paramomycin may be better for other tissues. The use of specific antibiotics and their analogues must be empirically determined for each target tissue. Other antibiotics that are commonly used with selectable marker genes in plants include hygromycin (FINER and MCMULLEN 1990), phleomycin (PEREZ et al. 1989) and gentamycin (HAJDUKIEWICZ et al. 1994).

The most commonly used herbicide resistance marker gene is the *bar* gene (WEEKS et al. 1993; BECKER et al. 1994) which is typically used with species of the Gramineae. The *bar* gene encodes phosphinothricin acetyl transferase, which inactivates phosphinothricin herbicides by acetylation. The *bar* gene can be used with bialaphos or glufosinate; again, selection of the proper herbicide must be empirically determined as with antibiotics. Other commonly used herbicide resistance genes impart resistance to glyphosate (ZHOU et al. 1995), sulfonyl ureas (SONGSTADT et al. 1996) and dalapon (BUCHANAN-WOLLASTON et al. 1992).

5.1.2 Reporter Genes

The most widely used reporter gene for plant transformation work is currently β-glucuronidase (GUS) (JEFFERSON 1987). For transient expression studies, plant tissue is typically evaluated for GUS expression 2–3 days after bombardment with the GUS gene by immersing the tissue in a solution containing 5-bromo-4-chloro-3-indolyl-β-D-glucuronic acid (x-gluc) for periods of 16–24h. Localized enzyme activity in the intact tissue can be visualized as blue spots or areas within the bombarded tissue. For clearer visualization of GUS-positive regions where chlorophyll interference is a problem, tissues can be transferred to 70%–100% ethanol for removal of chlorophyll after the blue coloration appears. Although GUS gene expression does not require specialized equipment for evaluation, a major disadvantage of using GUS is that histochemical staining results in tissue death. GUS activity in plant tissues can be quantified using a fluorometric assay with 4-methylumbelliferyl β-D-glucuronide (MUG) as the substrate (JEFFERSON 1987). Fluorometric analysis involves extraction of the GUS enzyme from the tissue and results in quantification of GUS expression. The real advantage of using GUS as a reporter gene is the simplicity of the GUS histochemical assay. Further, the blue GUS-expressing regions can easily be visualized without significant background in most bombarded plant tissues.

Although the GUS gene has proven to be very useful in monitoring transient expression in a number of species, nondestructive assays for transient expression would provide a significant advantage over a method resulting in tissue death. Plant tissues could be evaluated at various times following bombardment by reporter genes whose expression can be monitored without causing tissue damage. Such a system could greatly reduce the effort and time involved in selection of stable transformants.

The firefly luciferase gene is a nontoxic reporter gene that has been successfully used to monitor gene expression in plants (Ow et al. 1987; MILLAR et al. 1992). After the bombarded tissues are treated with luciferin (substrate) solution, luciferase activity is documented using either X-ray film (Ow et al. 1987) or specialized light-sensitive cameras (MICHELET and CHUA 1996) to detect a luminescent product. Use of X-ray film can be very inconvenient while low light cameras are somewhat costly and must be used in a dark room. Luciferase activity can also be assayed using extracts of plant tissue that are incubated with the luciferin substrate (GODON et al. 1993). Unfortunately, detection techniques for luciferase severely limit the flexibility of using this reporter gene.

Genes that encode transcription factors that regulate anthocyanin biosynthesis have also been used as reporter genes. For example, introduction of a chimeric R-gene under a constitutive promoter resulted in red anthocyanin pigmentation in maize tissues that could be easily visualized (LUDWIG et al. 1990). The R-gene has not received widespread use because high expression levels may impact plant regeneration and expression is strongly influenced by the genetic background of the target tissue.

Perhaps the most useful reporter gene in current use in plants is the green fluorescent protein (GFP) of the jellyfish (*Aequorea victoria*). The 238 amino acid native jellyfish GFP has three amino acids (Ser-65–Tyr-66–Gly-67) that cyclize to form a chromophore. The native protein has two excitation maxima by UV (396 nm) and blue (475 nm) light with respective emission maxima of 508 and 503 nm (CUBITT et al. 1995). Several modified versions of GFP have been synthesized with the aim of improving its performance as a plant reporter gene. One of the more effective of these versions has an introduced mutation so that the chromophore has threonine substituted for serine at position 65. This engineered "S65T" (which is "red-shifted") GFP has a single excitation peak when illuminated by blue light, and exhibits an elevated fluorescent signal, making detection much easier (CHIU et al. 1996). Wild-type GFP has been modified to alter codon usage, solubility, and/or stability of the protein in plants. These modifications, when combined with alterations in the chromophore, have yielded a very useful and sensitive reporter gene system (PANG et al. 1996; CHIU et al. 1996; HASELOFF and AMOS 1995; DAVIS and VIERSTRA 1996). Expression of GFP following particle bombardment has been reported in several species including Arabidopsis (CHIU et al. 1996; DAVIS and VIERSTRA 1996), onion (CHIU et al. 1996) and both corn and wheat (PANG et al. 1996). Depending on the target tissue, some red and yellow background fluorescence may be present which could potentially complicate GFP detection. This potential disadvantage may be overcome by use of an appropriately engineered GFP and/or suitable excitation and emission filters that reduce background fluorescence. While GFP has many potential benefits, it is a relatively new marker gene for plant transformation and the long-term effects of GFP expression on plant growth and development remain unclear.

5.1.3 Introns

The presence of introns between the promoter and coding region of a gene can significantly affect gene expression. This phenomenon was first described by CALLIS et al. (1987) and has subsequently been confirmed in several studies (FROMM et al. 1990; VASIL et al. 1991, 1992; NEHRA et al. 1994). Monocotyledonous species, in particular the Graminae, show significant enhancement of gene expression when using introns. Introns may act by stabilizing mRNAs, thus leading to greater amounts of translation products (LUEHRSEN and WALBOT 1991, 1994). Introduction of different introns between the cauliflower mosaic virus 35S promoter and GUS gene caused significant enhancement of transient gene expression in maize, although the degree of expression varied depending on the intron (VAIN et al. 1996). The same study also revealed that transient expression in bluegrass was unaffected by several introns that stimulated expression in maize, thus indicating that the strength of intron-containing gene constructs vary from species to species within the monocots. Of a number of chimeric GUS genes tested, the ubiquitin (*ubi 1*) intron of maize conferred the greatest activity in both maize and bluegrass in comparison with the intron-less 35S-GUS construct (VAIN et al. 1996). Another interesting observation in this study was that a dicot intron (from *chsA* of petunia) increased the level of GUS expression in both maize and bluegrass.

Although gene expression in many monocots is clearly affected by the presence of introns (VAIN et al. 1996), the effect of introns in genes introduced into dicots is somewhat less clear. In some studies, gene expression in dicots has increased slightly (NORRIS et al. 1993; LEON et al. 1991), whereas in other investigations, expression has not been stimulated by introns (VANCANNEYT et al. 1990; PASzKOWSKI et al. 1992). This suggests that the use of intron-containing constructs to enhance gene expression is more useful for monocots.

5.1.4 Promoter Analysis

The use of reporter genes such as GUS or GFP under control of strong promoters (e.g., CaMV 35S) allows for optimization of transformation efficiency of crop plants. Particle bombardment also provides a rapid and convenient means for evaluation of promoter strength, analysis of the function of different regions of promoters, and for examination of tissue specificity of gene expression. The promoter, either in its entirety or with various deletions/mutations, is fused to a reporter gene and introduced into the target tissue. This approach has successfully been used in a number of different species.

Quantitative analysis of a soybean glycinin promoter-GUS fusion in immature seeds and leaves by particle bombardment indicated seed specific GUS expression (IIDA et al. 1995) and subsequent analysis revealed possible regulatory regions within the promoter. MONTGOMERY et al. (1993) used promoter-luciferase fusions to identify regions of the tomato E4 promoter that are involved in ethylene responsiveness. The effects of various physiological conditions on gene expression have been determined by introducing suitable promoter-reporter fusions and observing the reporter gene activity as a result of regulatory factors. These include osmotic stress (ONDE et al. 1994; RAGHOTHAMA et al. 1993), hormones (KAO et al. 1996; RAGHOTHAMA et al. 1993; XU et al. 1996), temperature (WHITE et al. 1994) and light (BANSAL et al. 1992; KAO et al. 1996).

5.1.5 Codon Usage

The ability to introduce genes into plants from nonplant sources had led to the realization that genes from different organisms can be processed very differently with respect to their expression when compared with native plant genes (DIEHN et al. 1996). Unaltered DNA from nonplant sources may contain DNAs that are recognized by plants as signals for splicing or polyadenylation. In addition, nonplant DNAs may contain codons that are rare in plant genes. Occurrence of rare codons could actually reduce gene expression by becoming limiting during translation. DNA sequences can be modified to generate mRNA with the more abundant plant codons that are efficiently recognized and read by the translational machinery of plants. Modification and resynthesis of genes can be necessary to obtain high levels of expression of a nonplant gene. Modification of the native GFP gene from jellyfish was necessary to avoid missplicing of a cryptic intron in Arabidopsis (HASELOFF et al. 1997). Changes for codon usage resulted in a 300-fold increase in expression levels of a synthetic *Bt* gene (PERLAK et al. 1993).

5.2 Target Tissue

A variety of different plant tissues have been used as targets for particle bombardment-mediated transformation. Selection of the appropriate target tissue is dependent on the nature of the research. For rapid gene expression studies, various plasmid constructions can be introduced into different tissues and transient expression can be quickly analyzed to assess promoter activity without the production of stably transformed plants (IIDA et al. 1995). Almost any tissue can be used for transient expression studies as long as the cell wall is penetrable by the DNA-coated particles. The choice of the appropriate target tissue, physiological state of the plant material, and pre- and postbombardment treatments are critical to the successful regeneration of stable transformants.

5.2.1 Embryogenic Cultures

Embryogenic cell cultures have been very useful for the production of stably transformed plants. Particle bombardment of embryogenic maize and wheat cultures was used to generate the first transgenic plants of these species (FROMM et al.

1990; VAIN et al. 1993; VASIL et al. 1992). Embryogenic cultures provide a reliable source of tissue for optimization of bombardment conditions and can be used for year-round production of transgenic plants. An additional advantage of embryogenic cultures is that proliferation of transformed embryogenic tissue under selective conditions results in the production of solidly transformed embryogenic tissue, which subsequently gives rise to nonchimeric plants. However, the maintenance of long-term embryogenic cultures can be labor-intensive and time-consuming as the appropriate "type" of embryogenic callus or suspension cultured material must be selectively subcultured (REDWAY et al. 1990; VASIL et al. 1991, 1992). An additional complication is that long-term culture can result in abnormalities that may compromise the usefulness of the transgenic plant (RHODES et al. 1988). A 6-year-old embryogenic soybean line that was readily transformed gave rise to regenerated plants (from either transformed or nontransformed cultures) that were sterile (HADI et al. 1996). Interestingly, fertile soybean plants from short-term embryogenic cultures can be generated using similar methodologies, but the younger cultures are apparently not as receptive for stable transformation despite exhibiting high levels of transient expression of bombarded genes (Finer, unpublished).

As a result of the problems encountered with plants regenerated from long-term embryogenic cultures, transformation systems that involve rapid regeneration following particle bombardment have been developed (ALTPETER et al. 1996; BECKER et al. 1994; NEHRA et al. 1994). In wheat (BECKER et al. 1994) and maize (KOZIEL et al. 1993), immature embryos are either bombarded immediately after excision or cultured for a few days prior to bombardment. Although this approach has eliminated some of the problems of long-term cultures, some of the young target tissues are not very responsive to cell culture manipulations and the continual production of high quality immature embryos as initial explant material can be difficult.

5.2.2 Shoot Apical Meristems

One target tissue that most transformation scientists would like to be able to effectively target for production of stable transgenics is the shoot apical meristem. Theoretically, introduction of DNA into shoot meristem cells would result in the production of chimeric plants, where the transformed cells would directly give rise to germ-line tissue and the introduced DNA would be passed on to the progeny. This approach could potentially avoid or reduce tissue culture manipulations but there are many problems with shoot meristem bombardment. First, the target tissue within the shoot meristem can be a few cell layers deep and there are few laboratories that are able to accurately gauge and control the depth of particle penetration (MCCABE et al. 1988). With most particle guns, the particles penetrate only one or two cell layers (TAYLOR and VASIL 1991). Second, with particle bombardment, only a small proportion of the cells that receive the introduced genes and express transiently, yield stable events; meaning that a large number of meristems must be prepared. Lastly, the meristem itself is small and delicate and therefore somewhat tedious to prepare in large numbers. In spite of these apparent difficulties, particle bombardment of shoot meristematic cells has been used for the production of transgenic plants. It has not yet been possible to bombard and directly generate chimeric shoots that transmit the introduced gene to progeny (SAUTTER et al. 1991), but bombardment of shoot meristematic tissues followed by tissue culture multiplication has resulted in production of transgenics that transmit the introduced DNA to progeny (MCCABE et al. 1988).

In a direct comparison of bombardment of shoot tips and embryogenic cultures, SATO et al. (1993) bombarded shoot apices and embryogenic cultures of soybean. Although bombarded shoot tips exhibited limited GUS-expressing regions, GUS sectors were not observed in mature plants and the introduced DNA was not transmitted to progeny. In contrast, bombardment of embryogenic soybean suspensions in the same study yielded numerous GUS-positive clones that were regenerated into nonchimeric plants. Unfortunately, since the bombarded cultures were older, established lines, the plants regenerated from these cultures were sterile. Selection of clones from bombarded shoot apices can be time-consuming and labor intensive, but has resulted in the successful recovery of fertile transgenic plants (RUSSELL et al. 1993; LOWE et al. 1995). In addition, bombardment of meristematic tissue from shoot apices does not require the establishment of long-term callus/embryogenic cultures for bombardment of tissue and the possibility of culture-induced abnormalities is minimized.

In a variation of direct bombardment of the shoot meristem, BIDNEY et al. (1992) bombarded shoot apices of sunflower with particles alone (no DNA) and then inoculated the tissue with *Agrobacterium*. In this case, particle bombardment was used as a method to wound the target tissue for transformation via *Agrobacterium* rather than for direct DNA introduction itself. Although the resultant plants were chimeric for the introduced gene(s), it is important to note that many of the transformation events were germ-line and transgenic progeny were recovered.

5.2.3 Other Target Tissues

Particle bombardment can be used to introduce DNA into any target tissue that is penetrable by the DNA-coated particles. For stable transformation studies the target tissue should be regenerable, but for transient expression studies any tissue can be tested for expression of a reporter gene. In addition to embryogenic cultures and shoot tips, other tissues that have been subjected to particle bombardment include leaves (KLEIN et al. 1988), root sections (SEKI et al. 1991), stem sections (LOOPSTRA et al. 1992), pollen (TWELL et al. 1989; CLARK and SIMS 1994), styles (CLARK and SIMS 1994), cereal aleurone cells (KIM et al. 1992) and tassel primordia (DUPUIS and PACE 1993).

5.3 Tissue Treatment

In addition to choice of the appropriate target material, transformation efficiency can be affected by pre- and postbombardment treatments. One treatment that greatly influences gene expression is osmotic conditioning (FINER and MCMULLEN 1990,

1991; RUSSELL et al. 1992a; VAIN et al. 1993). Osmotic conditioning can be both pre- and postbombardment. Tissues are subjected to either partial drying in a laminar flow hood or cultured in a medium containing an osmotic agent such as mannitol and sorbitol at relatively high concentration ($\sim$0.4 M). Osmotic treatment can enhance both transient expression as well as significantly increase stable transformation. The osmotica may act by causing plasmolysis of the target cells that lessens or eliminates extrusion of the protoplasm in cells that are penetrated by particles (VAIN et al. 1993).

Other examples of media additives include the use of silver thiosulfate along with mannitol, which increased GUS expression in bombarded scutellar tissue of wheat, presumably by reducing the negative effects of wound ethylene (PERL et al. 1992). Inclusion of abscisic acid with an osmoticum (myo-inositol) increased transient GUS expression 1 day after bombardment of *Picea* embryogenic cultures (CLAPHAM et al. 1995).

The proper preculture of explant material is also beneficial for successful transformation (PERL et al. 1992; ALTPETER et al. 1996; WEEKS et al. 1993). Preculture refers to the culture of explant material for a few days prior to bombardment, rather than using freshly isolated explants. Preculture may condition the cells to make them more receptive to DNA uptake and integration. This area is often ignored although proper optimization of preculture conditions could result in significant improvements in transformation efficiency.

6 The Fate of the Introduced DNA(s)

Although particle bombardment has been used for DNA delivery to plant cells for some time, it is still unclear what happens to the majority of DNA that is introduced. Transient expression studies indicate that large numbers of cells can receive the particles resulting in approximately 10,000 GUS-positive cells per bombardment (FINER et al. 1992; VAIN et al. 1993). In cells that transiently express the introduced DNA, particles are most often seen either in or directly adjacent to the nucleus (YAMASHITA et al. 1991). Although the introduced DNA can express as soon as one hour after bombardment (Ponappa et al., unpublished), very few (if any) of the cells that show transient expression become stable transformation events. The majority of the delivered DNA therefore is either degraded or inactivated. In those cells that stably integrate the introduced DNA, the introduced gene may act as expected or express in a completely unpredictable manner. Studies of transgene expression patterns, taken together with the molecular analysis of the introduced DNA(s) do provide some information on the fate of DNA in bombarded plant cells.

6.1 Recombination

Molecular analysis of plant tissues transformed using particle bombardment give both simple and complex patterns that are typical for particle bombardment and other direct DNA uptake systems. Plasmid DNA which is introduced into plant cells via particle bombardment undergoes recombination with other plasmid molecules as well as with the host chromosomal DNA. An understanding of the results of the recombination process may help to possibly control this process in the future.

Recombination occurs whenever new gene arrangements are formed through exchange, elimination or insertion of DNA. The recombinational insertion of plasmid DNA into sites on the target plant chromosomal DNA is an integration or insertion event. Plasmid DNA can either insert cleanly into plant chromosomal DNA as a single entity or large linkages of plasmids can form and subsequently integrate (the full range of intermediate combinations is possible). Integration of a large chain of plasmids can result in a single insertion event with a high copy number of introduced plasmid DNA. With transformed plant tissue obtained via particle bombardment, a single integration event rarely results in the introduction of a single plasmid.

When large numbers of plasmids integrate into the same site, the plasmids may recombine with each other prior to integration into genomic DNA. If a mixture of two different plasmids is used, recombination can occur between like or unlike plasmids. Recombination and integration of unlike plasmids is called co-transformation. The recombination process itself can occur between the same region (homologous recombination) or different regions (illegitimate recombination) on plasmids. When inter-plasmid homologous recombination occurs, the plasmids seem to form end-to-end arrangements which appear as multiple copy, intense, unit-length bands when analyzed by Southern hybridization analysis (FINER and MCMULLEN 1990). When inter-plasmid illegitimate recombination occurs, Southern analysis results in single intensity fragments which are various lengths (HADI et al. 1996).

Although the exact fate of DNA introduced via particle bombardment is not clear, it appears that plasmid DNA molecules rearrange or recombine with each other to yield linkages of both whole plasmids and plasmid fragments. The linkages then insert into plant chromosomal DNA. The end result is an uncontrolled arrangement of whole plasmids and plasmid fragments, with one to over 100 copies of the introduced plasmid DNA, all integrated into one to five insertion sites. This end result is undesirable in many cases as the arrangement of plasmid DNA affects the expression of the introduced genes.

6.1.1 Position Effect

The expression of the introduced gene(s) is dependent on the nature of the DNA as well as the final physical arrangement of the foreign DNA in the plant genome. The influence that adjacent, native plant DNA has on expression of the introduced DNA is called "position effect" (DEAN et al. 1988; PEACH and VELTEN 1991). If the foreign DNA integrates in a region of DNA that is active, the introduced gene may be expressed at relatively high levels. If the introduction occurs in a region that is not transcriptionally active, gene expression may be reduced. Although the T-DNA from *Agrobacterium* is generally inserted into transcriptionally active regions (KONCZ et al. 1989; INGELBRECHT et al. 1991), DNA that is introduced via particle bombardment shows no known preference for insertion sites. Position effects have been suggested as one explanation for variation in gene expression in different transformed clones. Methylation of introduced DNA has been shown to occur in cases where the introduced gene shows minimal activity (MATZKE et al. 1989; PROLS and MEYER 1992), but the precise association of methylation with position effects is not clear.

6.1.2 Co-suppression

A reduction in expression of a gene resulting from the introduction of multiple copies of the gene is called "co-suppression" (NAPOLI et al. 1990). Co-suppression is useful in some situations when a reduction in expression of a native gene is desirable. Unfortunately, co-suppression is also observed as an undesirable product of particle bombardment-mediated transformation when multiple copies of the same gene are introduced. Transgenic plants that give inconsistent expression of the introduced gene(s) often contain multiple copies of the introduced gene (FINER and MCMULLEN 1991). Single-copy introductions are desirable and generally provide more consistent gene expression.

6.1.3 Scaffold Attachment Regions

To stabilize gene expression in transgenic plant tissue, scaffold attachment regions (SARs; also known as matrix attachment/associated regions, MARs) have shown some utility (ALLEN et al. 1993; BREYNE et al. 1992). SARs are regions of DNA which were isolated based on their ability to bind to the nuclear scaffold. For transformation, the SARs are ligated to the flanking regions of the gene of interest. They apparently act by stabilizing or normalizing gene expression rather than enhancing overall activity (BREYNE et al. 1992). SARs from yeast (ALLEN et al. 1993) and tobacco (BREYNE et al. 1992; ALLEN et al. 1996) have helped to stabilize gene expression in plants while a human and soybean SAR were nonfunctional (BREYNE et al. 1992). The level of stabilization of gene expression from SARs may be related to the affinity of that SAR towards the nuclear matrix of the target plant. Therefore, different SARs must be evaluated in different plants. Particle bombardment-mediated transformation efficiency could potentially be increased using SARs as SAR activity was found in a Transformation Booster Sequence isolated from petunia (GALLIANO et al. 1995). SARs appear to be especially useful in cases of high copy number DNA introductions, which would normally result in highly variable gene expression profiles.

6.2 Targeted Recombination Systems in Plants

In order to have better control on the expression of the introduced DNA(s), it would be beneficial to predict or direct the site of integration. This can be accomplished through either "agrolistic" transformation or homologous recombination. For agrolistic transformation (HANSEN and CHILTON 1996), plant tissues are bombarded with plasmids that contain some of the DNA excision components from the *Agrobacterium* gene transfer system. Approximately 20% of agrolistically transformed plant cells contained precisely defined DNA inserts that would be expected for *Agrobacterium*-mediated transformation. For targeting using homologous recombination, genes could potentially be placed in certain areas of the genome based on sequence homology between the introduced and the genomic DNA. Although targeted or homologous recombination is fairly well-characterized in some organisms (RADDING 1982; SMITH 1988; PETES et al. 1991), it has not been well-studied in higher plants. Very little is known about native plant recombinases (REISS et al. 1996; CERUTTI et al. 1992) with most recombination studies in plants relying on information and genes that are available from nonplant systems.

Two homologous recombination systems that have been evaluated in plants are the Cre-lox system from bacteriophage P1 (BAYLEY et al. 1992; ALBERT et al. 1995) and the FLP/FRT system from yeast (LYZNIK et al. 1993; BAR et al. 1996). In these systems, a target site containing specific DNA sequences must be introduced in the plant chromosomal DNA using transformation. Following activation with a specific recombinase, a second piece of DNA replaces the target DNA through homologous recombination. Although these nonplant recombination systems are not yet in wide use, there is much potential for targeted integration of introduced DNA once these systems are fine-tuned and better characterized.

7 Intellectual Property

With the spoils of gene transfer come protection of methodologies, DNAs and plant material used to generate those spoils. Particle bombardment is no different, having been patented in the United States by its inventor (SANFORD et al. 1990). The patenting of new methodologies is viewed by some as a way for greedy inventors to generate money for themselves. Others view patents as a way to protect inventions and promote commercialization of products, especially as the cost and risks associated with development of these products increase.

The owner of a patent has the legal right to prevent others from making, using, or selling inventions claimed by that patent. In the United States, construction of a particle bombardment device or use of the device to generate a product is covered under a series of patents exclusively licensed by E.I. duPont de Nemours (Wilmington, DE). Therefore construction of a particle bombardment device or use of the device to generate a product should not be done, unless a license is obtained. The lease or purchase of a particle acceleration system from BioRad (Hercules, CA), DuPont's exclusive licensee for the particle bombardment equipment, is usually accompanied by an agreement, defining the conditions under which the device can be operated. To use particle bombardment for transformation of crop plants, duPont should be contacted, while Sanford Scientific (Waterloo, NY) should be contacted for use with turf and ornamentals.

The intent of this section is not to provide a thorough review of the patent art but to make the reader aware that restrictions exist for using particle bombardment. A thorough review of the international patent art is suggested before practicing this technology.

Acknowledgements. Salaries and research support were provided by state and federal funds appropriated to OSU/OARDC and Kent State University. Mention of trademark or proprietary products does not constitute a guarantee or warranty of the product by OSU/OARDC and KSU and also does not imply approval to the exclusion of other products that may also be suitable. OARDC Journal Article No. 110-97.

References

Albert H, Dale EC, Lee E, Ow DW (1995) Site-specific integration of DNA into wild-type and mutant lox sites placed in the plant genome. Plant J 7:649–659

Allen GC, Hall G, Michalowski S, Newman W, Spiker S, Weissinger AK, Thompson WF (1996) High-level transgene expression in plant cells: effects of a strong scaffold attachment region from tobacco. Plant Cell 8:899–913

Allen GC, Hall GE, Childs LC, Weissinger AK, Spiker S, Thompson WF (1993) Scaffold attachment regions increase reporter gene expression in stably transformed plant cells. Plant Cell 5:603–613

Altpeter F, Vasil V, Srivastava V, Stoger E, Vasil IK (1996) Accelerated production of transgenic wheat (*Triticum aestivum* L.) plants. Plant Cell Rep 16:12–17

Bansal KC, Viret JF, Haley J, Khan BM, Schantz R, Bogorad L (1992) Transient expression from *cab-m1* and *rbcs-m3* promoter sequences is different in mesophyll and bundle sheath cells in maize leaves. Proc Natl Acad Sci USA 89:3654–3658

Bar M, Leshem B, Gilboa N, Gidoni D (1996) Visual characterization of recombination at FRT-gusA loci in transgenic tobacco mediated by constitutive expression of the native FLP recombinase. TAG 93:407–413

Bayley CC, Morgan M, Dale EC, Ow DW (1992) Exchange of gene activity in transgenic plants catalyzed by the Cre-lox site-specific recombination system. Plant Mol Biol 18:353–361

Becker D, Brettschneider R, Lorz H (1994) Fertile transgenic wheat from microprojectile bombardment of scutellar tissue. Plant J 5:299–307

Bidney D, Scelonge C, Martich J, Burrus M, Sims L, Huffman G (1992) Microprojectile bombardment of plant tissues increases transformation frequency by *Agrobacterium tumefaciens*. Plant Mol Biol 18:301–313

Bower R, Birch RG (1992) Transgenic sugarcane plants via microprojectile bombardment. Plant J 2:4099–416

Boynton JE, Gillham NW, Harris EH, Hosler JP, Johnson AM, Jones AR, Randolph-Anderson BL, Robertson D, Klein TM, Shark KB, Sanford JC (1988) Chloroplast transformation in Chlamydomonas with high velocity microprojectiles. Science 240:1524–1538

Breyne P, Montagu M van, Depicker A, Gheysen G (1992) Characterization of a plant scaffold attachment region in a DNA fragment that normalizes transgene expression in tobacco. Plant Cell 4:463–471

Buchanan-Wollaston V, Snape A, Cannon F (1992) A plant selectable marker gene based on the detoxification of the herbicide dalapon. Plant Cell Rep 11:627–631

Callis J, Fromm M, Walbot V (1987) Introns increase gene expression in cultured maize cells. Genes Dev 1:1183–1200

Cerutti H, Osman M, Grandoni P, Jagendorf AT (1992) A homolog of *E. coli* RecA protein in plastids of higher plants. Proc Natl Acad Sci USA 89:8068–8072

Chilton M-D, Drummond MJ, Merlo DJ, Sciaky D, Montoya AL, Gordon MP, Nester EW (1977) Stable incorporation of plasmid DNA into higher plant cells: the molecular basis of crown gall tumorigenesis. Cell 11:263–271

Chiu W-L, Niwa Y, Zeng W, Hirano T, Kobayashi H, Sheen J (1996) Engineered GFP as a vital reporter in plants. Curr Biol 6:325–330

Christou P (1994) Application to plants. In: Yang N-S, Christou P (eds) Particle bombardment technology for gene transfer. Oxford University Press, New York

Christou P, McCabe DE, Swain WF (1988) Stable transformation of soybean callus by DNA coated particles. Plant Physiol 87:671–674

Christou P, Swain WF, Yang N-S, McCabe DE (1989) Inheritance and expression of foreign genes in transgenic soybean plants. Proc Natl Acad Sci USA 86:7500–7504

Christou P, Ford T, Kofron M (1991) Production of transgenic rice (Oryza sativa L.) plants from agronomically important indica and japonica varieties via electric discharge particle acceleration of exogenous DNA into immature zygotic embryos. Bio Technology 9:957–962

Clapham D, Manders G, Yibrah HS, von Arnold S (1995) Enhancement of short- and medium-term expression of transgenes in embryogenic suspensions of Picea abies (L.) Karst. J Exp Bot 287:655–662

Clark KR, Sims TL (1994) The S-Ribonuclease gene of Petunia hybrida is expressed in nonstylar tissue, including immature anthers. Plant Physiol 106:25–36

Cubitt AB, Heim R, Adams SR, Boyd AE, Gross LA, Tsien RY (1995) Understanding, improving and using green fluorescent proteins. Trends Biochem Sci 20:448–455

Davis SJ, Vierstra RD (1996) Soluble derivatives of green fluorescent protein (GFP) for use in Arabidopsis thaliana. Weeds World 3:43–48

Dean C, Jones J, Favreau M, Dunsmuir P, Bedbrook J (1988) Influence of flanking sequences on variability in expression levels of an introduced gene in transgenic tobacco plants. Nucleic Acids Res 16:9267–9283

Diehn SH, DeRocher EJ, Green PJ (1996) Problems that can limit the expression of foreign genes in plants: lessons to be learned from Bt. toxin genes. In: Setlow JK (ed) Genetic engineering, principles and methods. Plenum, New York

Dupuis I, Pace GM (1993) Gene transfer to maize male reproductive structure by particle bombardment of tassel primordia. Plant Cell Rep 12:607–611

Finer JJ, McMullen MD (1990) Transformation of Cotton (Gossypium hirsutum) via particle bombardment. Plant Cell Rep 8:586–589

Finer JJ, McMullen MD (1991) Transformation of soybean via particle bombardment of embryogenic suspension culture tissue. In Vitro Cell Dev Biol 27P:175–182

Finer JJ, Vain P, Jones MW, McMullen MD (1992) Development of the particle inflow gun for DNA delivery to plant cells. Plant Cell Rep 11:232–238

Fromm ME, Morrish F, Armstrong C, Williams R, Klein TM (1990) Inheritance and expression of chimeric genes in the progeny of transgenic maize plants. Bio/Technology 8:833–839

Galliano H, Muller AE, Lucht JM, Meyer P (1995) The transformation booster sequence from Petunia hybrida is a retrotransposon derivative that binds to the nuclear scaffold. Mol Gen Genet 247:614–622

Godon C, Caboche M, Daniel-Vedele F (1993) Transient plant expression: a simple and reproducible method based on flowing particle gun. Biochimie 75:591–595

Gordon-Kamm WJ, Spencer TM, Mangano ML, Adams TR, Daines RJ, Start WG, O'Brien JV, Chambers SA, Adams WR, Willetts NG, Rice TB, Mackey CJ, Krueger RW, Kausch AP, Lemaux PG (1990) Transformation of Maize cells and regeneration of fertile transgenic plants. Plant Cell 2:603–618

Hadi MZ, McMullen MD, Finer JJ (1996) Transformation of 12 different plasmids into soybean via particle bombardment. Plant Cell Rep 15:500–505

Hajdukiewicz P, Svab Z, Maliga P (1994) The small, versatile pPZP family of Agrobacterium binary vectors for plant transformation. Plant Mol Biol 25:989–994

Hansen G, Chilton MD (1996) "Agrolistic" transformation of plant cells: integration of T-strands generated in planta. Proc Natl Acad Sci USA 93:14978–14983

Hansen G, Das A, Chilton MD (1994) Constitutive expression of the virulence genes improves the efficiency of plant transformation by Agrobacterium. Proc Natl Acad Sci USA 91:7603–7607

Haseloff J, Siemering KR, Prasher DC, Hodge S (1997) Removal of a cryptic intron and subcellular localization of green fluorescent protein are required to mark transgenic Arabidopsis plants. Proc Natl Acad Sci USA 94:2122–2127

Haseloff J, Amos B (1995) GFP in plants. Trends Genet 11:328–329

Hiei Y, Ohta S, Komari T, Kumashiro T (1994) Efficient transformation of rice (Oryza sativa L.) mediated by Agrobacterium and sequence analysis of the boundaries of the T-DNA. Plant J 6:271–282

Horsch RB, Fry JE, Hoffman NL, Eicholtz D, Rogers SG, Fraley RT (1985) A simple and general method for transferring genes into plants. Science 227:1229–1231

Iida A, Nagasawa A, Oeda K (1995) Positive and negative cis-regulatory elements in the soybean glycinin promoter identified by quantitative transient gene expression. Plant Cell Rep 14:539–544

Ingelbrecht I, Breyne P, Vancompernolle K, Jacobs A, VanMontagu M, Depicker A (1991) Transcriptional interference in transgenic plants. Gene 109:239–242

Ishida Y, Saito H, Ohta S, Hiei Y, Komari T, Kumashiro T (1996) High efficiency transformation of maize (*Zea mays* L.) mediated by *Agrobacterium tumefaciens*. Nat Biotechnol 14:745–750

Jefferson RA (1987) Assaying chimeric genes in plants. Plant Mol Biol Rep 5:387–405

Johnston SA, Anziano PQ, Shark K, Sanford JC, Butow RA (1988) Mitochondrial transformation in yeast by bombardment with microprojectiles. Science 240:1538–1541

Johnston SA, Riedy M, DeVit MJ, Sanford JC, McElligot S, Sanders Williams R (1991) Biolistic transformation of animal tissue. In Vitro Cell Dev Biol 27P:11–14

Kao CY, Cocciolone SM, Vasil IK, McCarty DR (1996) Localization and interaction of the *cis*-acting elements for abscisic acid, VIVIPAROUS1 and light activation of the C1 gene of maize. Plant Cell 8:1171–1179

Kim JK, Cao J, Wu R (1992) Regulation and interaction of multiple protein factors with the proximal promoter regions of a rice high pI α-amylase gene. Mol Gen Genet 232:383–393

Klein TM, Wolf ED, Wu R, Sanford JC (1987) High velocity microprojectiles for delivering nucleic acids into living cells. Nature 327:70–73

Klein TM, Harper EC, Svab Z, Sanford JC, Fromm ME, Maliga P (1988) Stable genetic transformation of intact *Nicotiana* cells by the particle bombardment process. Proc Natl Acad Sci USA 85:8502–8505

Koncz C, Martini N, Mayerhofer R, Koncz-Kalman Z, Körber H, Redei GP, Schell J (1989) High frequency T-DNA-mediated gene tagging in plants. Proc Natl Acad Sci USA 86:8467–8471

Koziel MG, Beland GL, Bowman C, Carozzi NB, Crenshaw R, Crossland L, Dawson J, Desai N, Hill M, Kadwell S (1993) Field performance of elite transgenic maize plants expressing an insecticidal protein derived from *Bacillus thuringiensis*. Bio/Technology 11:194–200

Leon P, Planckaert F, Walbot V (1991) Transient gene expression in protoplasts of *Phaseolus vulgaris* isolated from a cell suspension culture. Plant Physiol 95:968–972

Loopstra CA, Weissinger AK, Sederoff RR (1992) Transient gene expression in differentiating pine wood using microprojectile bombardment. Can J For Res 22:993–996

Lowe K, Bowen B, Hoerster G, Ross M, Bond D, Pierce D, Gordon-Kamm B (1995) Germline transformation of maize following manipulation of chimeric shoot meristems. Bio/Technology 13:677–682

Ludwig SR, Bowen B, Beach L, Wessler SR (1990) A regulatory gene as a novel visible marker for maize transformation. Science 247:449–450

Luehrsen KR, Walbot V (1991) Intron enhancement of gene expression and the splicing efficiency of introns in maize cells. Mol Gen Genet 225:81–93

Luehrsen KR, Walbot V (1994) Addition of A- and U-rich sequence increases the splicing efficiency of a deleted form of a maize intron. Plant Mol Biol 24:449–463

Lyznik LA, Mitchell JC, Hirayama L, Hodges TK (1993) Activity of yeast FLP recombinase in maize and rice protoplasts. Nucleic Acids Res 21:969–975

Matzke MA, Primig M, Trinovski J, Matzke AJM (1989) Reversible methylation and inactivation of marker genes in sequentially transformed tobacco plants. EMBO J 8:643–649

McCabe D, Christou P (1993) Direct DNA transfer using electrical discharge particle acceleration (Accell technology) Plant Cell Tissue Organ Cult 33:227–236

McCabe DE, Swain WF, Martinell BJ, Christou P (1988) Stable transformation of soybean (*Glycine max*) by particle acceleration. Bio/Technology 6:923–926

Michelet B, Chua N-H (1996) Improvement of Arabidopsis mutant screens based on luciferase imaging *in planta*. Plant Mol Biol Rep 14:320–329

Millar AJ, Short SR, Hiratsuka K, Chua NH, Kay SA (1992) Firefly luciferase as a reporter of regulated gene expression in higher plants. Plant Mol Biol Rep 10:324–337

Montgomery J, Goldman S, Deikman J, Margossian L, Fischer RL (1993) Identification of an ethylene-responsive region in the promoter of a fruit ripening gene. Proc Natl Acad Sci USA 90:5939–5943

Napoli C, Lemieux C, Jorgensen R (1990) Introduction of a chimeric chalcone synthase gene into petunia results in reversible co-suppression of homologous genes in transformation. Plant Cell 2:279–289

Nehra NS, Chibbar RN, Leung N, Caswell K, Mallard C, Steinhauer L, Baga M, Kartha KK (1994) Self-fertile transgenic wheat plants regenerated from isolated scutellar tissues following microprojectile bombardment with two distinct gene constructs. Plant J 5:285–297

Norris SR, Meyer SE, Callis J (1993) The intron of *Arabidopsis thaliana* polyubiquitin genes is conserved in location and is a quantitative determinant of chimeric gene expression. Plant Mol Biol 21:895–906

Onde S, Futers TS, Cuming AC (1994) Rapid analysis of an osmotic stress responsive promoter by transient expression. J Exp Bot 45:561–566

Ow DW, Jacobs JD, Howell SH (1987) Functional regions of the cauliflower mosaic virus 35 S RNA promoter determined by use of the firefly luciferase gene as a reporter of promoter activity. Proc Natl Acad Sci USA 84:4870–4874

Pang SZ, DeBoer DL, Wan Y, Ye G, Layton JG, Neher MK, Armstrong CL, Fry JE, Hinchee MAW, Fromm ME (1996) An improved green fluorescent protein gene as a vital marker in plants. Plant Physiol 112:893–900

Paszkowski J, Peterhans A, Bilang R, Filipowicz W (1992) Expression in transgenic tobacco of the bacterial neomycin phosphotransferase gene modified by intron insertions of various sizes. Plant Mol Biol 19:825–836

Peach C, Velten J (1991) Transgene expression variability (position effect) of CAT and GUS reporter genes driven by linked divergent T-DNA promoters. Plant Mol Biol 17:49 60

Perez P, Tiraby G, Kallerhoff J, Perret J (1989) Phleomycin resistance as a dominant selectable marker for plant cell transformation. Plant Mol Biol 13:365–373

Perl A, Kless H, Blumenthal A, Galili G, Galun E (1992) Improvement of plant regeneration and GUS expression in scutellar wheat calli by optimization of culture conditions and DNA-microprojectile delivery procedures. Mol Gen Genet 235:279 284

Perlak FJ, Stone TB, Miskopf YM, Petersen LJ, Parker GB, McPherson SA, Wyman J, Love S, Reed G, Biever D, Fischoff DA (1993) Genetically improved potatoes: protection from damage by Colorado potato beetles. Plant Mol Biol 22:313–321

Petes TD, Malone RE, Symington LS (1991) Recombination in yeast. In: Broach JR, Pringle JR, Jones EW (ed) The molecular and cellular biology of yeast saccharomyces: genome dynamics, protein synthesis and energetics. Cold Spring Harbor Laboratory, Cold Spring Harbor, pp. 407 521

Prols F, Meyer P (1992) The methylation patterns of chromosomal integration regions influence gene activity of transferred DNA in *Petunia hybrida*. Plant J 2:465 475

Radding CM (1982) Homologous pairing and strand exchange in genetic recombination. Annu Rev Genet 16:405–437

Raghothama KG, Liu D, Nelson DE, Hasegawa PM, Bressan RA (1993) Analysis of an osmotically regulated pathogenesis-related osmotin gene promoter. Plant Mol Biol 23:1117 1128

Redway FA, Vasil V, Vasil IK (1990) Characterization of wheat (*Triticum aestivum* L.) embryogenic cell suspension cultures. Plant Cell Rep 8:714 717

Reiss B, Klemm M, Kosak H, Schell J (1996) RecA protein stimulates homologous recombination in plants. Proc Natl Acad Sci USA 93:3094–3098

Rhodes CA Lowe KS Ruby KL (1988) Plant regeneration from protoplasts isolated from embryogenic maize cell cultures. Biotechnology 6:56–60

Russell JA, Roy MK, Sanford JC (1992a) Physical trauma and tungsten toxicity reduce the efficiency of biolistic transformation. Plant Physiol 98:1050–1056

Russell JA, Roy MK, Sanford JC (1992b) Major improvements in biolistic transformation of suspension-cultured tobacco cells. In Vitro Cell Dev Biol 28P:97 105

Russell DR, Wallace KM, Bathe JH, Martinell BJ, McCabe DE (1993) Stable transformation of *Phaseolus vulgaris* via electric discharge mediated particle acceleration. Plant Cell Rep 12:165 169

Sanford JC (1988) The biolistic process. Trends Biotechnol 6:299–302

Sanford JC, Klein TM, Wolf ED, Allen N (1987) Delivery of substances into cells and tissues using a particle bombardment process. J Part Sci Tech 5:27–37

Sanford JC, Wolf ED, Allen NK (1990) Method for transporting substances into living cells and tissues and apparatus therefor. US Patent #4945050

Sanford JC, Devit MJ, Russell JA, Smith FD, Harpening PR, Roy MK, Johnston SA (1991) An improved helium driven biolistic device. Technique 3:3–16

Sato S, Newell C, Kolacz K, Tredo L, Finer JJ, Hinchee M (1993) Stable transformation via particle bombardment in two different soybean regeneration systems. Plant Cell Rep 12:408 413

Sautter C, Waldner H, Neuhaus-Url G, Galli A, Niehaus G, Potrykus I (1991) Micro-targeting: high efficiency gene transfer using a novel approach for the acceleration of micro-particles. Bio Technology 9:1080–1085

Seki M, Shigemoto N, Komeda Y, Imamura J, Yamada Y, Morikawa H (1991) Transgenic *Arabidopsis thaliana* plants obtained by particle-bombardment-mediated transformation. Appl Microbiol Biotechnol 36:228–230

Smith FD, Harpending PR, Sanford JC (1992) Biolistic transformation of prokaryotes: factors that affect biolistic transformation of very small cells. J Gen Microbiol 138:239 248

Smith GR (1988) Homologous recombination in prokaryotes. Microbiol Rev 52:1 28

Somers DA, Rines HW, Gu W, Kaeppler HF, Bushnell WR (1992) Fertile, transgenic oat plants. Bio/Technology 10:1589–1594

Songstad DD, Armstrong CL, Petersen WL, Hairston B, Hinchee MAW (1996) Production of transgenic maize plants and progeny by bombardment of Hi-II immature embryos. In Vitro Cell Dev Biol Plant 32:179–183

Stachel SE, Messens E, Van Montagu M, Zambryski P (1985) Identification of the signal molecules produced by wounded plant cells which activate the T-DNA transfer process in *Agrobacterium tumefaciens*. Nature 318:624–629

Sundaram P, Xiao W, Brandsma JL (1996) Particle mediated delivery of recombinant expression vectors to rabbit skin induces high titered polyclonal antisera (and circumvents purification of a protein immunogen) Nucleic Acids Res 24:1375–1377

Taylor MG, Vasil IK (1991) Histology of, and physical factors affecting, transient GUS expression in pearl millet (*Pennisetum glaucum* (L.) R Br) embryos following microprojectile bombardment. Plant Cell Rep 10:120–125

Tinland B, Hohn B (1995) Recombination between prokaryotic and eukaryotic DNA: integration of *Agrobacterium tumefaciens* T-DNA into the plant genome. In: Setlow JK (ed) Genetic engineering. Plenum, New York, pp 209–229

Toffaletti DL, Rude TH, Johnston SA, Durack DT, Perfect JR (1993) Gene Transfer in *Cryptococcus neoformans* by use of biolistic delivery of DNA. J Bacteriol 175:1405–1411

Twell D, Klein TM, Fromm ME, McCormick S (1989) Transient expression of chimeric genes delivered into pollen by microprojectile bombardment. Plant Physiol 91:1270–1274

Vain P, Finer KR, Engler DE, Pratt RC, Finer JJ (1996) Intron-mediated enhancement of gene expression in maize (*Zea mays* L.) and bluegrass (*Poa pratensis* L.) Plant Cell Rep 15:489–494

Vain P, McMullen MD, Finer JJ (1993) Osmotic treatment enhances particle bombardment-mediated transient and stable transformation of maize. Plant Cell Rep 12:84–88

Vancanneyt G, Schmidt R, O'Connor-Sanchez A, Willmitzer L, Rocha-Sosa M (1990) Construction of an intron-containing marker gene: splicing of the intron in transgenic plants and its use in monitoring early events in Agrobacterium-mediated plant transformation. Mol Gen Genet 220:245–250

Vasil V, Brown SM, Re D, Fromm ME, Vasil IK (1991) Stably transformed callus lines from microprojectile bombardment of cell suspension cultures of wheat. Bio/Technology 9:743–747

Vasil V, Castillo AM, Fromm ME, Vasil IK (1992) Herbicide resistant fertile transgenic wheat obtained by microprojectile bombardment of regenerable embryogenic callus. Biotechnology 10:667–674

Weeks TJ, Anderson OD, Blechl AE (1993) Rapid production of multiple independent lines of fertile transgenic wheat (*Triticum aestivum*) Plant Physiol 102:1077–1084

White TC, Simmonds D, Donaldson P, Singh J (1994) Regulation of BN115, a low-temperature-responsive gene from winter *Brassica napus*. Plant Physiol 106:917–928

Xu R, Goldman S, Coupe S, Deikman J (1996) Ethylene control of E4 transcription during tomato fruit ripening involves two cooperative *cis* elements. Plant Mol Biol 31:1117–1127

Yamashita T, Iida A, Morikawa H (1991) Evidence that more than 90% of β-glucuronidase-expressing cells after particle bombardment directly receive the foreign gene in their nucleus. Plant Physiol 97:829–831

Zhou H, Arrowsmith JW, Fromm ME, Hironaka CM, Taylor ML, Rodriguez D, Pajeau ME, Brown SM, Santino CG, Fry JE (1995) Glyphosphate-tolerant CP4 and GOX genes as a selectable marker in wheat transformation. Plant Cell Rep 15:159–163

Plant Viral Vectors Based on Tobamoviruses

V. Yusibov[1], S. Shivprasad[2], T.H. Turpen[3], W. Dawson[2], and H. Koprowski[1]

1 Introduction

Viruses are ubiquitous in plants, particularly those associated with agriculture. Plant viruses have a range of features that extend from detrimental to potentially beneficial. Economic loses to agriculture caused by viral infections led to the development of genetic systems that allow manipulation of the virus to manage plant diseases; however, these genetic systems have also led to the development of viruses as beneficial tools, exploiting the ability of the small plus-sense single-stranded RNA viruses that commonly infect higher plants to rapidly amplify virus-related RNAs and produce large amounts of proteins. As early as 1983 Siegel described a strategy for the development of virus-based vectors to express foreign genes in plants, although at that time the technology for manipulation of viruses through cDNA cloning was not yet available, and the issue was controversial (SIEGEL 1983;

[1] Biotechnology Foundation Laboratories, Thomas Jefferson University, Philadelphia, PA 19107, USA
[2] University of Florida, Lake Alfred, FL 33850, USA
[3] Biosource Technologies Inc., Vacaville, CA 95688, USA

VAN VLOTEN-DOTING 1985). Since that time, in vitro genetic systems have been developed for viruses in different taxonomic groups and several of these have been manipulated to transiently express foreign genes (for review, see KEARNEY et al. 1995).

High-level expression of the introduced genes and the rapid accumulation of appropriate products that are easily purified from the host plant are the primary characteristics that make plant viruses well suited as transient expression vectors. Moreover, plant viruses generally have a wide host range that allows expression of a gene in different plant species using the same vector construct. The considerable body of knowledge about virus biology, genetics and regulatory sequences needed for effective expression of foreign genes in plants has enabled refinement in developing optimal virus vectors, including the selection of virus-based vectors that have minimal disruptive effects on the plant host.

Efforts to manipulate plant viruses in vitro began with DNA viruses such as geminiviruses and caulimoviruses. The latter were considered potential candidates as plant viral vectors (GRONENBORN et al. 1981; BRISSON et al. 1984; DE-ZOETEN et al. 1989), but to date have found only marginal use. Geminivirus vectors, based on single-stranded DNA viruses that replicate in the nucleus, also have considerable potential as tools to examine plants and are still under development. The vast majority of viruses that infect plants are single-stranded RNA viruses that replicate in the cytoplasm. Many of these viruses replicate to very high levels in plants and a few have been used to express foreign genes (SCHOLTHOF and SCHOLTHOF 1996). One such virus is tobacco mosaic virus, the focus of this review.

2 Development of Plant Tobamovirus Based Expression Vectors

Tobacco mosaic virus (TMV) and its relatives in the tobamovirus group have several advantages for consideration as transient expression vectors. One is their long history of experimentation. TMV was the first virus to be recognized and purified, and the first virus for which virion structure and the coat protein sequence were determined. TMV has easily manipulable cDNA clones that provide highly infectious in vitro produced transcripts that are easily transmitted to host plants. With no efficient natural vector, TMV also represents a reduced bio-hazard as a recombinant virus in the field. Moreover, TMV has a wide host range which enables the analysis of newly engineered hybrid virus constructs in a variety of plant species. One of TMV's hosts, tobacco, is among the highest biomass and protein-producing crops, and under optimal conditions, 2000kg tobacco protein can be produced per acre per year (TURPEN et al. 1995). Finally, TMV virions can easily be purified in large quantities by relatively simple procedures and equipment. The design of vectors that fuse the desired products to the virion, with engineered cleavage sites, provides an effective means of obtaining large quantities of purified products.

TMV is a single-stranded plus-sense RNA virus with a genome of 6395 nucleotides with only four open reading frames (GOELET et al. 1982). The 126-kDa protein and 183-kDa readthrough protein are translated from the 5' end of the genomic RNA. These two proteins, together with putative host proteins, form the replicase complex. The 30-kDa movement protein and 17.5-kDa capsid protein are translated from 3'-coterminal subgenomic mRNAs produced during the replication of the viral RNA (HUNTER et al. 1976; BEACHY and ZAITLIN 1977). The movement protein is necessary for the virus to pass through the plasmodesmatal opening between cells to support cell-to-cell movement. The capsid protein is the single structural unit protecting the RNA in virions and is required for long-distance movement throughout the plant.

The coat protein of TMV is one of the most accumulated proteins in plants, accounting for up to 10% of the dry weight of infected leaves (SIEGEL et al. 1978; FRASER 1987). The structure of the coat protein is known (NAMBA et al. 1989; BUTLER et al. 1992), and substantial information is available on the effect of amino acid substitutions on the formation of tighter or looser subunit interactions and virion formation (CULVER et al. 1994, 1995). Thus, substitution of the viral coat protein with a foreign protein resulted in one of the first hybrid constructs (TAKAMATSU et al. 1987; Dawson et al. 1988). Substitution of coat protein with chloramphenicol acetyl transferase (CAT) resulted in a free-RNA virus that generated high CAT activity; however, the amount of CAT was substantially less relative to the TMV coat protein produced in plants infected with wild-type virus. In addition, the deficiency in coat protein production limited the virus to locally infected tissue. For most applications, an effective vector must move throughout the plant.

DAWSON et al. (1989) created an "extra-gene" vector by cloning the CAT gene under the control of an extra coat protein subgenomic promoter of TMV into the entire TMV genome. The hybrid virus produced an additional (third) subgenomic mRNA that was translated into active CAT and replicated efficiently. The inserted sequences, however, were rapidly deleted from the virus, with most of the virus in the upper leaves being wild-type. The rapid deletion of the CAT sequences from this vector apparently occurred by the homologous recombination of the repeated subgenomic promoter sequences. Previously, it was shown that TMV hybrids with repeated coat protein genes always deleted an extra copy of coat protein gene during early replication (BECK and DAWSON 1990); however, a similar construct where coat protein genes from both TMV and Odontoglossom ringspot virus (ORSV) were expressed, was extraordinarily stable. This virus remained infectious and did not delete any of the coat protein sequences during several passages from plant to plant (CULVER et al. 1995). These results suggested that a vector with a functionally duplicated region should utilize heterologous instead of homologous repeated sequences.

Based on the above information, DONSON et al. (1991) developed a TMV-based hybrid plant viral expression vector, TB2, which had a heterologous coat protein

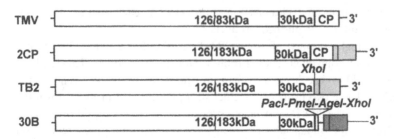

Fig. 1. Genomic organization TMV, 2CP, TB2, and 30B. *Light shading*, ORSV CP sequences; *dark shading*, TMGMV U5 CP and 3'UTR sequences. In TB2 the start condon of coat protein is mutated to ACG and in 30B to AGA. The coat protein subgenomic promoter of CP extends up to nt 5756 in 30B. The TMV U1 pseudoknots precede the TMGMV U5 sequences

gene and subgenomic promoter from ORSV (Fig. 1). The open reading frames (ORFs) for neomycin phosphotransferase or dihydrofolate reductase were controlled by the subgenomic promoter of the TMV coat protein, and coat protein expression was directed by the subgenomic promoter from ORSV. TB2 efficiently produced the foreign proteins and stably maintained the inserted sequences during systemic infection of *N. benthamiana*.

Although it had been argued that RNA viruses were unsuitable as vectors to produce foreign proteins in plants because of the low fidelity of RNA polymerases (VAN VLOTEN-DOTING et al. 1985), experimentation has shown virus-based vectors to be unexpectedly stable. KEARNEY et al. (1993) examined the accumulation of replicase errors in foreign genes inserted into TB2, which were thought to be nonselected neutral sequences that would accumulate errors. The TB2 vector continuously replicated during a 6-month period through ten passages from plant to plant via systemically infected tissue in controlled laboratory conditions. RNA isolated and assessed for possible errors of RNA-dependent RNA polymerase during virus replication revealed no mutations in viral RNA during the ten passages. The replication error frequency was less than the error frequency of the in vitro reverse transcriptase-polymerase chain reaction (RT-PCR); however, every recombinant virus eventually loses its insert. Often, virus from the upper leaves can be propagated through a succession of plants, but gradually the population becomes more as wild-type TMV. The stability of the constructs containing different foreign genes can differ substantially; however, instability is a major advantage for production of recombinant proteins in the field. None of the constructs of TMV-based transient expression vectors are sufficiently stable to survive in nature and each construct quickly reverts to a crippled wild-type-like virus. Thus, the hybrid viruses do not survive in nature. CASPAR and HOLT (1996) developed a similar "extra-gene" TMV-based vector, 4GD-PL, with a heterologous coat protein subgenomic mRNA promoter from tomato mild green mosaic virus (TMGMV), but with a homologous coat protein ORF and nontranslated 3' sequences. This vector efficiently expresses foreign proteins systemically throughout plants as demonstrated by the expression of green fluorescent protein (GFP) in a TMV-based vector in *Nicotiana benthamiana*.

3 Improved TMV Vectors and New Applications

Mapping of the TMV coat protein subgenomic mRNA promoter demonstrated that the vector TB2 did not contain the entire promoter for optimal expression of the foreign gene. The optimal promoter, which produced maximum amounts of subgenomic message, mapped into the coat protein ORF (Chapman and Lewandowski, unpublished results). Based on the previous demonstration that proximity of genes to the 3'-untranslated region of the genome increases the efficiency of their translation (CULVER et al. 1993), the position effect in TB2 was examined using a double coat protein hybrid virus containing the coat protein genes of TMV and ORSV under the control of their natural subgenomic promoters, respectively (Fig. 1). Although approximately equal amounts of subgenomic RNA were generated for each coat protein gene, the RNA predominantly translated was from the gene located nearest the 3' untranslated region; however, when the so-called pseudoknot region was duplicated between the two coat protein genes, the amounts of protein expressed from both genes were approximately equal (Chapman and Lewandowski, unpublished results). In independent studies, the 3' terminus carrying three pseudoknots and a tRNA-like structures (VAN BELKUM et al. 1985; PLEIJ et al. 1987; MANS et al. 1991) was shown to influence translation of heterologous messages (LEATHERS et al. 1993). These results suggested that expression of internal genes could be increased by adjacent placement of an internal set of pseudoknot sequences. Based on the above information, a new vector, 30B, was designed. In this vector, the start codon (AUG) of the coat protein was mutated to AGA, and a series of cloning sites (PacI, PmeI, AgeI, and XhoI) were engineered 40 nucleotides downstream of the original AUG to provide a full-size subgenomic RNA promoter. This was followed by the TMV pseudoknot sequences from the 3' untranslated region (UTR), and the TMV U5 subgenomic promoter, coat protein ORF, and 3'-UTR (Fig. 1). The U5 coat protein sequence was used to replace the ORSV gene because of its support of efficient long-distance movement in tobacco. The vector can be monitored easily with the reporter gene, cycle 3 mutant of GFP from *Aequorea victoria*. This vector is very stable in *N. benthamiana*, with GFP being visible even 2 months after inoculation, and supports high levels of GFP production.

4 Application of Viral Vectors

Genetically modified plant viruses are clearly powerful tools with a variety of applications. The opportunity to accumulate specific recombinant proteins in whole plants as a cost-effective source of therapeutic reagents for human and animal disease conditions has immediate potential in health care applications. Success will depend on safety, efficacy and potency of the products used in either a purified form or in more crude oral or topical formulations. Most analysis of recombinant protein expression in plants do not progress beyond the measurement of accumulated cross-reacting immunologic material expressed as a percentage of the total soluble protein. In a compatible viral/host combination, many recombinant protein products expressed in plants can accumulate to 1%–5% of the total soluble protein, provided the protein is localized to an appropriate subcellular compartment. This cross-reacting immunologic material must be purified and structure/activity relationships further established before pharmaceutical applications can be realized. Viral transfection tools are necessary to rapidly optimize expression strategies for different proteins.

5 Expression of Full-Length Proteins

5.1 α-Galactosidase A

Many enzymes can be readily assayed for catalytic properties and therefore provide a convenient measure of recombinant product fidelity across heterologous hosts. The ribosome-inactivating protein α-trichosanthin was the first recombinant enzyme of pharmaceutical interest to be purified from plants (KUMAGAI et al. 1993). In this case, the specific activity of the enzyme purified from virus-transfected leaves appeared identical to that of the enzyme from the native source (roots of *Trichosanthes kirowlii*); however, this was a plant-derived natural product and the measurement of activity was indirect, based on the relative inhibition of translation in vitro. The significance of these observations was recently extended through expression of a human lysosomal enzyme, α-galactosidase A (Turpen and Pogue, unpublished results). Human α-galactosidase A was expressed in leaves with a tobamovirus vector because of the need for an inexpensive source of enzyme to treat a hereditary storage disorder known as Fabry's disease (DESNICK et al. 1995). This native human cDNA also provided a more challenging opportunity to assess the capacity of virus-transfected leaves to accumulate active products, because the mature enzyme requires disulfide-mediated folding, N-linked glycan site occupancy, and homodimerization of 50-kD subunits for activity. Several constructs were tested, containing active signal-peptide sequences for entry into the endoplasmic reticulum. These constructs accumulated similar levels of cross-reacting immunologic material in transfected leaves (~2% of total soluble protein); however, only after deletion of a portion of a putative carboxy-terminal propeptide (QUINN et al. 1987) was significant enzymatic activity observed. This activity was recovered from plant apoplast, a network of intercellular fluid, cell wall and extracellular matrix material located between adjacent cells of the leaf. Because leaf cells do not normally secrete significant quantities of protein, α-galactosidase was by far the predominant protein (>30%) in the intercellular fluid (Turpen and Pogue, unpublished results). As measured using a fluorescent substrate analog, enzyme purified from the leaf intercellular fluid had a specific activity equal to that of the enzyme purified from human tissues, from cultures of Chinese hamster ovary and from insect cells transfected with baculovirus. This example dramatically illustrates the significance of accurate analytical measurements of product quality. Some glycoforms of this enzyme were found to be over two orders of magnitude more active than others based on equivalent mass as measured in Western analysis. This principle applies to many other recombinant proteins of significance in immunology and other applications.

5.2 Single-Chain Antibodies for Passive Immunotherapy

Therapeutic proteins are now being used to treat several types of cancer. In the case of B cell lymphoma, whole antibodies and antibody fragments have been used to induce cellular and humoral immunity to tumor cells in patient-specific active vaccination programs (Hsu et al. 1997). Although clinical trials have shown very promising results, the purification of whole antibody from a tumor cell hybridoma limits both the rate and quantity of therapeutic protein production (CARROLL et al. 1986). Single-chain antibodies are an attractive alternative to whole antibody therapies, and are of equivalent biological activity in the 38C13 mouse model of B cell lymphoma (HAKIM et al. 1996).

To determine whether single-chain antibodies produced in plants are of equal biological activity to that of whole antibody, McCormick et al. (submitted) tested purified single-chain antibody of the 38C13 tumor-specific variable regions for therapeutic efficacy in mice challenged with B cell tumor cells. Expression of the 38C13 single-chain antibody (38C13-scFv) was directed to the apoplastic space of leaves transfected with tobamovirus vectors. On average, recombinant protein accumulated to 1%–2% of the total soluble protein approximately 2 weeks post-inoculation with about 30% of this product obtained in an initial intercellular fluid extraction. Using an antibody specific for the 38C13 antigen-binding site, ELISA analysis confirmed that the plant-made 38C13-scFv adopts the correct antigen binding domain in solution. 38C13-scFv protein was purified to near homogeneity for animal vaccinations from crude plant extracts by antibody affinity chromatography. Syngeneic C3H mice were injected intraperitoneally (IP) every 2 weeks for three vaccinations with 15µg 38C13-scFv protein made in plants and compared to those vaccinated with either 50µg 38C13Ig-KLH conjugate or with no protein as controls. Specific anti-38C13 antibodies were detected at microgram milliliter levels in the sera of all protein-vaccinated animals, but not in the nonprotein control animals. Two weeks after the last vaccination, animals were challenged IP with 38C13 tumor cells. At day 21 post-tumor challenge (PTC), all control animals had died with palpable abdominal tumors. At day 21, all but one 38C13IgKLH vaccinated animals were alive and healthy, with subsequent mortalities observed at days 25–43 PTC. At 40d PTC, 7/10 38C13IgKLH vaccinated animals and 6/10 38C13-scFv vaccinated animals remained tumor-free. There were no statistical differences between whole antibody, or plant-made scFv survival curves by log-rank analysis. Because this plant derived scFv can confer protective immunity to tumor challenge similar to the native 38C13 Ig, these results offer hope that cumbersome, time consuming and expensive hybridoma antibody production for human lymphoma therapies may someday be replaced by inexpensive, rapid, and effective scFv therapeutic proteins produced in plants.

5.3 Cytokines

Cytokines produced by plants exhibit high levels of biological activity as measured by nitric oxide release and virus protection cellular assays. Various types of interferon have been shown to accumulate to levels up to 1% of total soluble protein in leaves transfected with tobamovirus vectors. Such interferon preparations from plants also exhibit similar specific activities to material generated in *E. coli*, yeast or mammalian systems (Pogue and Hanley, personal communication).

During the rapid burst of viral protein synthesis, no saturation of several key posttranslational processing steps occurs, nor is there induction of host components that damage the products, accelerate turnover, or prevent purification. While the plant leaf does not secrete significant quantities of protein into the intercellular fluid, it has the potential to do so through the default pathway of protein sorting, which relies on bulk membrane vesicle traffic. Large-scale industrial pilot processes for recovery of recombinant protein from either leaf homogenates or the intercellular fluid of field-grown tobacco have recently been implemented, thus removing a final barrier to the introduction of bulk-purified pharmaceutical proteins from plants (Biosource Technologies, Bioprocessing Group, Owensboro, KY).

A series of other hybrid tobamovirus vectors have been generated which appear to be almost identical to 30B. These vectors are hybrids of TMV and other tobamoviruses with similar configurations, yet they vary considerably in levels of expression and stability. Some replicate to high levels, but produce minimal amounts of foreign protein. Some immediately delete the inserted sequences. Clearly, the construction of an efficient transient expression vector is much more complicated than simply inserting an extra subgenomic mRNA promoter fused to your favorite ORF.

6 Fusion Vectors

6.1 Subunit Vaccines

A variety of medically important antigens have been expressed in transgenic plants, including hepatitis B surface antigen (Mason et al. 1992), *E. coli* heat-labile enterotoxin (Haq et al. 1995), rabies virus glycoprotein (McGarvey et al. 1995), and Norwalk virus capsid protein (Mason et al. 1996). Each of these plant-produced proteins was shown to induce serum IgG specific for the original viral antigen in mice. Oral feeding of mice with potato tubers expressing either the *E. coli* heat-labile enterotoxin or the Norwalk virus capsid protein induced mucosal IgA and serum IgG specific for the respective pathogen. This response was greatly enhanced by the addition of cholera toxin as adjuvant. Mice fed on freeze-dried tomatoes expressing the rabies glycoprotein (without cholera toxin) mounted a virus-specific mucosal IgA response but no detectable serum IgG (Dr. P. McGarvey, personnel communication). In pigs, oral administration of transgenic potato tubers expressing the S protein of transmissible gastroenteritis virus resulted in reduction of morbidity (46%) and mortality (40%) upon subsequent challenge with infectious virus (WELTER et al. 1996). Clearly, antigens expressed in plants and delivered orally can elicit an antibody response, suggesting the promise of vaccines based on antigens produced in plants. However, the level of antigenic protein produced by transgenic plants is relatively low, suggesting the need for new approaches to express foreign proteins in plants.

An alternative approach to the use of transgenic plants is the use of engineered plant viruses for the production of foreign proteins or peptides, specifically, the engineering of virus coat proteins to function as carrier molecules for antigenic peptides. Such carrier proteins have the potential to self-assemble and form recombinant virus particles, displaying the desired antigenic epitopes on their surfaces. The first plant virus protein to be used as a carrier molecule for antigenic epitopes from other sources was the TMV coat protein (HAYNES et al. 1986). Since then, antigenic determinants of several pathogens have been successfully fused to the TMV coat protein and expressed in virally infected plants (HAMAMOTO et al. 1993; TURPEN et al. 1995; FITCHEN et al. 1995). The antigenic peptides were expressed either as a readthrough fusion, where up to 5% of the resulting TMV coat protein is recombinant (TURPEN et al. 1995) or as a translational fusion where every molecule of the coat protein carries the epitope (TURPEN et al. 1995; FITCHEN et al. 1995). The antigenic determinants produced using the TMV coat protein have been shown to be functional and able to stimulate the generation of specific antibodies in immunized animals (TURPEN et al. 1995; FITCHEN et al. 1995). Coat proteins of several other plant viruses have also been studied for their ability to serve as a carrier molecule, including cowpea mosaic virus (USHA et al. 1993; PORTA et al. 1994; MCLAIN et al. 1995; DALSGAARD et al. 1997), tomato bushy stunt virus (JOELSON et al. 1997) and alfalfa mosaic virus (YUSIBOV et al. 1997; MODELSKA et al. 1998). MCLAIN et al. (1995) showed that antigens produced in plants as a result of infection with engineered cowpea mosaic viruses can elicit specific antibodies when injected into mice. Using the coat protein of alfalfa mosaic virus as a carrier molecule, YUSIBOV et al. (1997) were able to stimulate neutralizing antibody production against rabies and HIV in mice. Mice immunized with recombinant virus containing antigenic determinants of rabies virus were protected against lethal challenge with the virus (MODELSKA et al. 1998). Recently, cowpea mosaic virus was used to induce protective immunity in mink against mink enteritis virus (DALSGAARD et al. 1997). The coat proteins of plant viruses clearly have great potential as a system for the expression and delivery of a variety of antigenic determinants.

6.2 Antimicrobial Peptides

Many researchers rely on synthetic peptides for experimental purposes, but are of the opinion that only recombinant products have the potential to be cost-effective on a clinical scale. Many different antimicrobial peptides are synthesized by plants and animals as pre-formed defense compounds against pathogens (RAO 1995). In general, they appear to function by perturbing target membranes. This mechanism of action presents obvious problems for engineering biological production systems.

There are several categories of peptides with distinct molecular structures. For example, the mammalian defensins are a relatively large family of structurally homologous compounds isolated from humans, rabbits, guinea pigs and rats. The defensins are highly cationic peptides, 29–35 amino acids in length and contain conserved disulfide-linked cysteine residues necessary for biological activity. They are found stored in granules of leukocytes and macrophages and have potent microbicidal activity against a wide variety of human pathogens, including fungi, gram-positive and gram-negative bacteria, spirochetes, mycobacteria, protozoans, and some enveloped viruses (MURPHY et al. 1993). Indolicidin is a linear antimicrobial molecule isolated from bovine neutrophil granules. This novel tridecapeptide is very tryptophan- and proline-rich and has potent bactericidal activity (SELSTED et al. 1992). Indolicidin was synthesized in tobacco plants as a TMV coat protein fusion product at a yield of over 300mg fusion protein per kilogram of leaves using the readthrough mechanism (Reinl and Turpen, unpublished results). Because protocols exist for protein folding of these molecules in vitro after cleavage of the fusion protein with cyaonogen bromide, it is sufficient to confirm the peptide quality by mass spectroscopy (MS). The matrix-assisted laser desorption ionization time-of-flight MS spectra of the indolicin fusion protein showed no unanticipated covalent modifications of protein structure in peptides derived from the cytosol of leaf tissues. Very similar results were obtained for the defensin NP-1. Therefore, plant viral expression vectors will enable cost-effective ribosomal-based synthesis of antimicrobial peptides using field grown plants as novel bioreactors.

7 Overexpression or Silencing of Host Genes

Perhaps one of the most important potential uses of transient expression vectors is in the functional analysis of unknown genes. There are numerous techniques such as screening libraries, yeast two-hybrid assays, chromosome walking, antibody screens and others which can identify unknown genes. The function of unknown genes can be screened using virus-based vectors in two ways. One is to overexpress the gene product. KUMAGAI et al. (1995) used a TMV-based vector to overexpress the phytoene synthase ORF in tomato plants, resulting in bright orange leaves with accumulated high levels of phytoene. Those investigators also expressed a partial ORF encoding phytoene desaturase in both the sense and antisense orientation in the TMV-based vector. Those experiments demonstrated the second use of virus vectors in screening for gene function, i.e., the newly developing leaves turned white due to inhibition of carotenoid biosynthesis. Similar experiments have recently been reported using a PVX-based expression vector (BAULCOMBE et al. 1997).

8 Summary

The potential of plant virus-based transient expression vectors is substantial. One objective is the production of large quantities of foreign peptides or proteins. At least one commercial group (Biosource Technologies) is producing large quantities of product in the field, has built factories to process truck-loads of material and soon expects to market virus-generated products. In the laboratory, large amounts of protein have been produced for structural or biochemical analyses. An important aspect of producing large amounts of a protein or peptide is to make the product easily purifiable. This has been done by attaching peptides or proteins to easily purified units such as virion particles or by exporting proteins to the apoplast so that purification begins with a highly enriched product. For plant molecular biology, virus-based vectors have been useful in identifying previously unknown genes by overexpression or silencing or by expression in different genotypes. Also, foreign peptides fused to virions are being used as immunogens for development of antisera for experimental use or as injected or edible vaccines for medical use. As with liposomes and microcapsules, plant cells and plant viruses are also expected to provide natural protection for the passage of antigen through the gastrointestinal tract. Perhaps the greatest advantage of plant virus-based transient expression vectors is their host, plants. For the production of large amounts of commercial products, plants are one of the most economical and productive sources of biomass. They also present the advantages of lack of contamination with animal pathogens, relative ease of genetic manipulation and the presence eukaryotic protein modification machinery.

References

Baulcombe D, Davenport G, Angell S, English J, Ratcliff F, Ruiz T, Voinnet O, Hamilton A (1997) American Phytopath. Soc Aug 9–13, Rochester
Beachy RN, Zaitlin M (1977) Characterization and in vitro translation of the RNAs from less than full-length, virus related, nucleoprotein rods present in tobacco mosaic virus preparations. Virology 81:160–169
Beck DL, Dawson WO (1990) Deletion of repeated sequences from tobacco mosaic virus mutants with two coat protein genes. Virology 177:462–469

Brisson N, Paszkowski J, Penswick JR, Gronenborn B, Potrykus I, Hohn T (1984) Expression of a bacterial gene in plants by using a viral vector. Nature 310:511–514

Butler PJG, Bloomer AC, Finch JT (1992) Direct visualization of the structure of the "20S" aggregate of coat protein of tobacco mosaic virus. The "disk" is the major structure at pH\7.0 and the proto-helix at lower pH. J Mol Biol 224:381–394

Carroll WL, Thielemans K, Dilley J, Levy R (1986) Mouse x human heterohybridomas as fusion partners with human B cell tumors. J Immunol Methods 89:61–72

Casper SJ, Holt CA (1996) Expression of the green fluorescent gene from a tobacco mosaic virus-based vector. Gene 173:69–73

Culver JN, Dawson WO, Plonk K, Stubbs G (1995) Site-directed mutagenesis confirms the involvement of carboxylate groups in the disassembly of tobacco mosaic virus. Virology 206:724–730

Culver JN, Lehto K, Close SM, Hilf ME, Dawson WO (1993) Genomic position affects the expression of tobacco mosaic virus movement and coat protein genes. Proc Natl Acad Sci USA 90:2055–2059

Culver JN, Stubbs G, Dawson WO (1994) Structure-function relationship between tobacco mosaic virus coat protein and hypersensitivity in *Nicotiana sylvestris*. J Mol Biol 242:130–138

Dalsgaard K, Uttenthal A, Jones TD, Xu F, Merryweather A, Hamilton WDO, Langeveld JPM, Boshuizen RS, Kamstrup S, Lomonossoff GP, Porta C, Vela C, Casal JI, Meloen RH, Rodgers PB (1997) Plant-derived vaccine protects target animals against a viral disease. Nat Biotechnolnol 15:248–252

Dawson WO, Lewandowski DJ, Hilf ME, Bubrick P, Raffo AJ, Shaw JJ, Grantham GL, Desjardins PR (1989) A tobacco mosaic virus-hybrid expresses and loses an added gene. Virology 173:285–292

Dawson WO, Bubrick P, Grantham GL (1988) Modifications of the tobacco mosaic virus coat protein gene affecting replication, movement, and symptomology. Phytopathology 78:783–789

Desnick RJ, Ioannou YA, Eng CM (1995) α-Galactosidase A deficiency: Fabry disease. In: Scriver CR, Beaudet AL, Sly WS, Valle D (eds) The metabolic bases of inherited diseases. McGraw-Hill, New York, pp 2741–2784

de Zoeten GA, Penswick JR, Horisberger MA, Ahl P, Schultze M, Hohn T (1989) The expression, localization, and effect of a human interferon in plants. Virology 172:213–222

Donson J, Kearney CM, Hilf ME, Dawson WO (1991) Systemic expression of a bacterial gene by a tobacco mosaic virus-based vector. Proc Natl Acad Sci USA 88:7204–7208

Fitchen J, Beachy RN, Hein MB (1995) Plant virus expressing hybrid coat protein with added murine epitope elicits autoantibody response. Vaccine 13:1051–1057

Fraser RSS (1987) Biochemistry of virus-infected plants. Research Studies, Letchworth, pp 1–7

Goelet P, Lomonossoff GP, Butler PJG, Akam ME, Gait MJ, Karn J (1982) Nucleotide sequence of tobacco mosaic virus RNA. Proc Natl Acad Sci USA 79:5818–5822

Gronenborn B, Gardner RC, Schaefer S, Shepherd RJ (1981) Propagation of foreign DNA in plants using cauliflower mosaic virus as vector. Nature 294:773–776

Hakim I, Levy S, Levy R (1996) A nine-amino acid peptide from IL-1β augments antitumor immune responses induced by protein and DNA vaccines. J Immunol 157:5503–5511

Hamamoto H, Sugiyama Y, Nakagawa N, Hashida E, Matsunaga Y, Takemoto S, Watanabe Y, Okada Y (1993) A new tobacco mosaic virus vector and its use for the systemic production of angiotensin-I-converting enzyme inhibitor in transgenic tobacco and tomato. Biotechnology 11:930–932

Haq TA, Mason H, Clements JD, Arntzen CJ (1995) Oral immunization with a recombinant bacterial antigen produced in transgenic plants. Science 268:714–716

Haynes JR, Cunningham J, von Seefried A, Lennick M, Garvin RT, Shen SH (1986) Development of genetically engineered candidate polio vaccine employing the self-assembling properties of tobacco mosaic virus coat protein. Biotechnology 4:637–641

Hsu FJ, Caspar CB, Czerwinski D, Kwak LW, Liles TM, Syrengelas A, Taidi-Laskowski B, Levy R (1997) Tumor-specific, idiotype vaccines in the treatment of patients with B-cell lymphoma-long-term results of a clinical trial. Blood 89:3129–3135

Hunter T, Hunt T, Knowland J, Zimmern D (1976) Messenger RNA for the coat protein of tobacco mosaic virus. Nature 260:759–764

Joelson T, Akerblom L, Oxefelt L, Strandberg B, Tomenius K, Morris TJ (1997) Presentation of a foreign peptide on the surface of tomato bushy stunt virus. J Gen Virol 78:1213–1217

Kearney CM Donson J, Jones GE, Dawson WO (1993) Low level of genetic drift in foreign sequences replicating in an RNA virus in plants. Virology 192:11–17

Kearney CM, Chapman S, Turpen TH, Dawson WO (1995) High levels of gene expression in plants using RNA viruses as transient expression vectors. Plant Molecular Biology Manual. Kluwer Academic, Dordrecht

Kumagai MH, Donson J, Della-Cioppa G, Harvey D, Hanley K, Grill LK (1995) Cytoplasmic inhibition of carotenoid biosynthesis with virus-derived RNA. Proc Natl Acad Sci USA 92:1679 1683

Kumagai MH, Turrpen TH, Weinzettl N, Della-Cioppa G, Turpen AM, Donson J, Hilf ME, Grantham GL, Dawson WO, Chow TP, Piatak Jr. M, Grill LK (1993) Rapid high-level expression of biologically active α-trichosanthin in transfected plants by an RNA viral vector. Proc Natl Acad Sci USA 90:427–430

Leathers V, Tanguay R, Kobayashi M, Gallie DR (1993) A phylogenetically conserved sequence within viral 3' untranslated pseudoknots regulates translation. Mol Cell Biol 13:5331 5347

Mans RMW, Pleij CWA, Bosch L (1991) tRNA-like structures. Structure function and evolutionary significance. Eur. J Biochem 201:303 324

Mason HS, Lam DM-K, Arntzen CJ (1992) Expression of hepatitis B surface antigen in transgenic plants. Proc Natl Acad Sci USA 89:11745 11749

Mason HS, Ball JM, Shi JJ, Jiang X, Estes MK, Arntzen CJ (1996) Expression of Norwalk virus capsid protein in transgenic tobacco and potato and its oral immunogenicity in mice. Proc Natl Acad Sci USA 93:5335 5340

McCormick AA, Kumagai MK, Hanley K, Turpen TH, Hakim I, Grill LK, Tusé D, Levy S, Levy R (submitted) Rapid production of specific vaccines for lymphoma by expression of the tumor-derived single chain Fv epitopes in tobacco plants

McGarvey PB, Hammond J, Dienelt MM, Hooper DC, Fu ZF, Dietzschold B, Koprowski H, Michaels FH (1995) Expression of the rabies virus glycoprotein in transgenic tomatoes. Biotechnology 13:1484 1487

McLain L, Porta C, Lomonossoff GP, Durrani Z, Dimmock NJ (1995) Human immunodeficiency virus type I-neutralizing antibodies raised to a glycoprotein 41 peptide expressed on the surface of a plant virus. AIDS Res Hum Retroviruses 11:327 334

Modelska A, Dietzschold B, Flyesh N, Fu ZF, Steplewski K, Hooper DC, Koprowski H, Yusibov V (1998) Immunization against rabies with plant-derived antigen. Proc Natl Acad Sci USA 95:2481 2485

Namba K, Pattaneyek R, Stubbs G (1989) Visualization of protein-nucleic acid interactions in a virus: refined structure of intact tobacco mosaic virus at 2.9 A resolution by X-ray fiber diffraction. J Mol Biol 208:7583–7588

Pleij CWA, Abrahams JP, van Belkum A, Rietveld K, Bosch L (1987) The spatial folding of the 3' noncoding region of aminoacylatable plant viral RNAs. In: Brinton M, Rueckert F (eds) Positive strand RNA viruses. Liss, New York, pp 299 316

Porta C, Spall VE, Loveland J, Johnson JE, Barker PJ, Lomonossoff GP (1994) Development of cowpea mosaic virus as a high-yielding system for the presentation of foreign peptides. Virology 202:949 955

Quinn M, Hantzopoulos P, Fidanza V, Calhoun DH (1987) A genomic clone containing the promoter for the gene encoding the human lysosomal enzyme, α-galactosidase A. Gene 58:177 188

Ramakrishnan U, Rohll JB, Spall VE, Shanks M, Maule AJ, Johnson JE, Lomonossoff GP (1993) Expression of an animal virus antigenic site on the surface of a plant virus particle. Virology 197:366 374

Scholthof HB, Morris TJ, Jackson AO (1993) The capsid protein gene of tomato bushy stunt virus is dispensable for systemic movement and can be replaced for localized expression of foreign genes. Mol Plant Microbe Interact 6:309 322

Scholthof HB, Scholthof K-BG (1996) Plant virus gene vectors for transient expression of foreign proteins in plants. Annu Rev Phytopathol 34:299 323

Siegel A (1983) RNA viruses as cloning vehicles. Phytopathology 73:775

Siegel A, Hari V, Kolacz K (1978) The effect of tobacco mosaic virus infection and virus-specific protein synthesis in protoplasts. Virology 85:494

Takamatsu N, Ishikawa N, Meshi T, Okada Y (1987) Expression of bacterial chloramphenicol acetyltransferase gene in tobacco plants mediated by TMV-RNA. EMBO J 6:307

Turpen TH, Reinl S, Charoenvit Y, Hoffman SL, Fallarme V, Grill LK (1995) Malarial epitopes expressed on the surface of recombinant tobacco mosaic virus. Biotechnology 13:53 57

Usha R, Rohll JB, Spall VE, Shanks M, Maule AJ, Johnson JE, Lomonossoff GP 1993. Expression of an animal virus antigenic site on the surface of a plant virus particle. Virology 197:366 374

van Belkum A, Abraham JP, Pleij CWA, Bosch L (1985) Five pseudoknots are present at the 204 nucleotide long 3' noncoding region of tobacco mosaic virus RNA. Nucleic Acids Res 13:7673 7686

van Vloten-Doting L, Bol JF, Cornelissen B (1985) Plant-virus-based vectors for gene transfer will be of limited use because of the high error frequency during viral RNA synthesis. Plant Mol Biol 4:323 326

Welter LM, Mason HM, Lu W, Lam DM-K, Welter MW (1996) Effective immunization of piglets with transgenic potato plants expressing a truncated TGEV S protein. Vaccines: new technologies and applications. Cambridge Healthtech Institute

Yusibov V, Modelska A, Steplewski K, Agadjanyan M, Weiner D, Hooper C, Koprowski H (1997) Antigens produced in plants by infection with chimeric plant viruses immunize against rabies virus and HIV-1. Proc Natl Acad Sci USA 94:5784–5788

Transgenic Plants for Therapeutic Proteins: Linking Upstream and Downstream Strategies

C.L. Cramer[1], J.G. Boothe[2], and K.K. Oishi[3]

1 Introduction

With the new knowledge generated through the Human Genome Project and related biomedical research comes a potential revolution in drug development strategies. One of the most direct applications of this knowledge will be highly specialized recombinant protein-based therapeutics. Recombinant drugs such as human erythropoietin (EPO), tissue plasminogen activator (tPA), and Cerezyme™ (glucocerebrosidase) are currently on the market and many other recombinant proteins are in various stages of human clinical trials. Commercial production of

---

[1] CropTech Corp., Virginia Tech Corporate Research Center, Blacksburg, VA, 24060 and Department of Plant Pathology, Physiology and Weed Science, Fralin Biotechnology Center, Virginia Polytechnic Institute and State University, Blacksburg, VA 24061-0346, USA
[2] Sem BioSys Genetics Inc., Suite 204, 609 14th St. N.W., Calgary, Alberta, T2N, Canada
[3] CropTech Corp., Virginia Tech Corporate Research Center, Blacksburg, VA, 24060, USA

these proteins utilizes fermentation (primarily *E. coli* and yeast) and mammalian cell systems (e.g., Chinese hamster ovary cells), the major expression systems adopted by the well established biotechnology companies. However, these expression systems have significant limitations. Bacteria cannot perform the complex posttranslational modifications required for bioactivity of many human proteins and high-level expression often leads to accumulation of insoluble protein aggregates. While mammalian cell cultures perform the required protein modifications, low transgene expression levels, instability of selected cell lines, and the difficulties and high expense of scale-up are often limiting or severely impact cost. Thus, there remains significant opportunity for alternative expression systems that address these limitations and cost issues to compete in the protein therapeutics market. In fact, development of more cost-effective protein bioproduction systems may be critical in translating the discoveries of genomics and medical research into widely available and affordable treatments and cures. Recent advances in the area of transgenics – the use of genetically engineered plants and animals for bioproduction – indicate great promise as effective protein factories. The fact that recombinant proteins from both transgenic animals and transgenic plants are now in clinical trials demonstrates significant progress toward commercialization of these technologies.

For any particular target protein, selection of a recombinant system will depend on the characteristics of the desired protein product, the volume needs (size of the market), and market-driven cost constraints (reviewed by PEN 1996). Transgenic plants have some remarkable features that make them particularly well suited for cost-effective bioproduction of proteins for pharmaceutical uses. These include: (a) low production costs, (b) reduced time to market, (c) unlimited supply, (d) eukaryotic protein processing, and (e) safety. Cost advantages are based not only on the low cost of biomass production, but also costs associated with research and development, germplasm scale-up (e.g., imagine the infrastructure investment of tripling the capacity of one's aseptic fermentation or mammalian cell production facility compared to tripling one's acreage for plant growth), and reduced requirements for quality assurance testing for exclusion of human pathogenic agents (reviewed in OWEN and PEN 1996). Plant-based strategies also have advantages in the pace at which feasibility testing can be done and R & D successes can be scaled up and brought to market. For example, a tobacco plant goes from seed to next generation seed in three months and produces up to a million seed per plant. Scaling-up to hundred or thousands of acres is very rapid.

Many of the therapeutic proteins of interest require complex posttranslational processing and/or oligomerization for bioactivity or appropriate targeting following administration to patients. There appears to be remarkable conservation of these protein processing steps between plants and animals such that the majority of human proteins that have been produced in plants (see Table 1) show significant structural, biochemical and functional equivalency to proteins from humans or animal cell cultures. In cases where certain modification steps are lacking or differ in plants (e.g., glycan composition, discussed further below), strategies to introduce appropriate animal protein processing enzymes or modify the plant processing

Table 1. Fidelity of plant-based production of human (or other animal) proteins

Transgene product	Potential use disease target	Plant host	Structural integrity	Functional activity	Reference
Serum proteins					
Haemoglobin (α and β)	Blood substitute	Tobacco	Yes (dimer)	Yes (O_2, CO_2 binding)	DIERYCK et al. 1997
Human serum albumin	Blood extender	Potato	Yes	Not tested	SIJMONS et al. 1990
Protein C	Anticoagulant	Tobacco	Most processing steps performed	Not tested	CRAMER et al. 1996a
Cytokines lymphokines					
α-Interferon	Viral protection, anti-cancer	Rice	Yes	Yes (viral resistance assay)	ZHU et al. 1994
γ-Interferon	Phagocyte activator	Tobacco	Yes (glycans[a])	Yes (in vitro assay)	GRILL 1997
CM-CSF	Leukopoiesis in bone marrow transplants	Tobacco	Yes (glycans[a])	Yes (growth stimul. of TF-1 cells)	GANZ et al. 1996
Epidermal growth factor	Mitogen	Tobacco	CRIM[b]	Not tested	HIGO et al. 1993
Trout growth factor	Mitogen	Tobacco	Yes (glycans[a])	Not tested	BOSCH et al. 1994
Erythropoietin	Mitogen, blood cells	Tobacco cells	Yes (glycans[a])	Not tested	MATSUMOTO et al. 1995
Lysosomal enzymes					
α-Galactosidase	Fabry disease	Tobacco	Yes (glycans[a])	Yes (enzyme act.)	GRILL 1997
Glucocerebrosidase	Gaucher disease	Tobacco	Yes (glycans[a])	Yes (enzyme act.)	CRAMER et al. 1996b
Other proteins					
Casein	Neutriceutical	Potato	CRIM[b]	Not tested	CHONG et al. 1997
Hirudin	Anticoagulant	Canola	Yes	Yes (thrombin inh.)	PARMENTER et al. 1995
NP1 defensin	Antibiotic	Tobacco	CRIM[b]	Yes (antibiotic act.)	GRILL 1997
Glutamate decarboxylase	Diabetes	Tobacco	CRIM[b]	Yes (mouse model)	MA et al. 1997

[a] Proteins were glycosylated but the glycan composition may differ from those produced in humans
[b] Detected as cross-reactive immundetected material by western immunoblots or ELISA

machinery are greatly facilitated by the ease of plant transformation and the broad experience in transgenic approaches to modifying plant metabolism through over-expression and antisense strategies. In fact, plants may be the only system capable of efficient production of certain human proteins such as growth regulators and cell cycle inhibitors which would negatively impact either the transgenic animal or animal cell culture in which they are expressed.

Perhaps the most important advantage of plants, which is emerging in the aftermath of the recent "mad cow disease" scare, involves product safety. The biopharmaceutical industry is now faced with the possibility of product validation and quality assurances that demonstrate purity not only from known human pathogens such as HIV but also from unknown or poorly characterized agents such as the prions responsible for bovine spongioform encephalopathy and the related Creutzfeld-Jakob disease (ROBERTS and JAMES 1996; VAUGHAN 1996). Plants do not serve as hosts for blood- or animal tissue-borne human pathogens. In addition, plant-based production and purification can be executed without the use of any animal-derived products. Clearly purity, efficacy and quality control issues similar to production of any biopharmaceutical will need to be addressed (see MIELE 1997). However, plant-based bioproduction should realize substantial savings as a human- and animal-source-free production system.

The list of complex human proteins and animal, viral and bacterial proteins of medical value that have been successfully expressed in plants is growing rapidly (reviewed in OWEN and PEN 1996). In addition to disease antigens (vaccines) and antibodies discussed in other chapters of this volume, transgenic plants have been used to synthesize a number of complex serum proteins, cytokines, growth regulators, anticoagulants, antibiotics, and lysosomal enzymes (see Table 1). Most of these proteins appear fully functional and structurally comparable to the analogous proteins produced in animal cell cultures or in humans. Thus, plants have clearly passed the initial test of feasibility – they are capable of producing bioactive human proteins of pharmaceutical value. In addition, the first transgenic plant-synthesized products (a tobacco-derived antibody targeting gum disease and a potato-derived edible vaccine candidate) have reached initial human trials – a significant benchmark toward commercialization. However, as we move from feasibility studies to commercial bioproduction, issues of transgene expression levels, product processing and stability, biomass and extraction scale-up, purification, and quality control become paramount. These longer-term goals have inspired the development of novel transgene expression systems that incorporate components targeting product abundance, product recovery, and regulatory acceptance into the initial transgene design. In this review, we will discuss key issues that impact the choice and utility of plant-based production systems for biopharmaceuticals. We will highlight several production strategies that stress the importance of linking "upstream" steps in genetic engineering and expression strategies with "downstream" issues of extraction, purification, and yield. These systems are designed to separate biomass production from transgenic protein production and to directly manipulate the timing, tissue and subcellular localization of the product to enhance yield, protein stability, and ease of recovery and purification.

2 Plant-Based Biopharmaceutical Production: Issues and Answers

The majority of examples demonstrating bioproduction of potential therapeutic proteins in plants shown in Table 2 have used model plant species that are easy to genetically engineer (e.g., tobacco, potato) and the "strong, constitutive" 35S

Table 2. Transgene expression strategies and recombinant protein yield

Transgene product	Plant host	Promoter	Expression strategy, tissue	Production levels (% of soluble protein)	Reference
Serum proteins					
Haemoglobin (α and β)	Tobacco	35S[a]	Seed, root	0.05%	DIERYCK et al. 1997
Human serum albumin	Potato	35S[a]	Constitutive, leaf	0.02%	SIJMONS et al. 1990
Protein C	Tobacco	35S	Constitutive, leaf	0.002%	CRAMER et al. 1996a
Cytokines lymphokines					
α-Interferon	Rice	Pl'	Constitutive, leaf	Not reported	ZHU et al. 1994
γ-Interferon	Tobacco	NA	Geneware viral-infected leaf	1%	GRILL 1997
CM-CSF	Tobacco	Rice glutelin	Seed-specific	Not reported	GANZ et al. 1996
Epidermal growth factor	Tobacco	35S	Constitutive, leaf	0.001%	HIGO et al. 1993
Lysosomal enzymes					
α-Galactosidase	Tobacco	NA	Geneware/leaf	12.1mg kg tissue	GRILL 1997
Glucocerebrosidase	Tobacco	MeGA	Post-harvest induced leaf tissues	1%–10%	CRAMER et al. 1996b
Viral or bacterial antigens					
E. coli enterotoxin B	Potato	35S[a]	Microtuber	0.003%	HAQ et al. 1995
Cholera toxin	Tobacco	35S[a]	Constitutive, leaf	Not reported	HEIN et al. 1996
Hepatitis B surf. antigen	Tobacco	35S[a]	Constitutive, leaf	0.007%	MASON et al. 1992
Other proteins					
Hirudin	Canola	Oleosin	Seed-specific	0.3% (seed protein)	PARMENTER et al. 1995
Antibodies	Tobacco	35S	Constitutive	0.01%–5%	Rev. in OWEN and PEN 1996
α-Trichosantin	Tobacco	NA	Geneware/leaf	4%–5%	KUMAGAI et al. 1993
Glutamate decarboxylase	Tobacco	35S[a]	Constitutive, leaf	0.4%	MA et al. 1997
	Potato	35S[a]	Tuber	0.4%	MA et al. 1997

[a] Modified 35S promoter containing enhancer duplication and/or leader sequences (translational enhancer) from the tobacco etch virus or alfalfa mosaic virus

promoter derived from the cauliflower mosaic virus. However, as plant biotechnology moves from demonstrating feasibility toward commercialization of protein products, many other issues come into play in selecting host species, expression strategies, target tissues, and extraction/purification protocols. These choices must take into account not only the production of the particular protein of interest but issues of recovery, purity, production/purification costs, reproducibility, supply continuity, quality control, and regulatory assessment.

2.1 Selection of Crop Species

While certain features such as low production costs and high biomass capacity are common to all plant-based expression systems, other factors may strongly influence the choice of one plant species or expression strategy over another for the production of a specific foreign protein. In selecting a particular species it is important to consider how readily it can be manipulated to produce a stable transgenic line, the tissue and subcellular compartment best suited for stable expression of the heterologous protein, and the availability of methods for the efficient harvesting and initial processing of the plant material. Included in the first consideration are factors such as the amenability to transformation and regeneration of whole plants, generation time, and tractability to controlled genetic crossing. All of these factors significantly impact upon the time and resources required for product development. Plant transformation technologies are highlighted in other chapters (Hansen and Chitton and Finer et al., this volume) and have been recently reviewed (LINDSEY 1996) and are therefore not discussed in detail here. The remaining two considerations deal mainly with product biocomparability (bioactivity, conformation, efficacy) and recovery. Because infrastructure and methods for the harvest and processing of the major crop species already exist, whenever possible these are the species of choice. The tissue and subcellular compartment of expression determines protein processing capabilities, stability of the product and the ease with which it can be recovered.

Tobacco remains the easiest plant to genetically engineer and is widely used to test suitability of plant-based systems for bioproduction of recombinant proteins (see Table 2). Although tobacco is considered a regional crop and relatively labor intensive, at least three plant-based biotech companies are targeting tobacco for biopharmaceutical production (CropTech Corp; BioSource Technologies, Inc. and Planet Biotechnology). In addition to being easily engineered, tobacco is an excellent biomass producer (in excess of 40 tons leaf fresh weight/acre based on multiple mowings per season) and prolific seed producer (up to one million seeds produced per plant), thus hastening the time in which a product can be scaled up and brought to market.

Several companies are developing production strategies involving transgene product accumulation in seeds, an organ designed to accumulate and store protein reserves (see Sect. 2.2). Companies targeting seed-based production using canola, corn or soybeans include Sem BioSys Genetics, Agracetus (USA), Mogen International (the Netherlands), and Plantzyme (the Netherlands). Applied Phytologics (API, Davis, CA) is using transgenic rice and barley seed but is producing and recovering recombinant proteins during seed germination in a process analogous to malting. Other crops being developed for biopharmaceutical protein or vaccine production include alfalfa, banana, potato, and tomato.

2.2 Choice of Tissue

In order to obtain maximum yields, the plant species selected must concentrate biomass in the organ or tissue where the foreign protein is expressed. The diversity among different species in this respect means that a variety of options are available including leaves, vegetative storage organs (e.g. tubers) and seeds. The tissue chosen should be compatible with the desired protein, enabling correct processing, stable accumulation and, whenever possible, efficient recovery. Many human therapeutic proteins require extensive processing for full activity, involving transport through the cellular endomembrane system. Functional lysosomal enzymes (CRAMER et al. 1996b) and mammalian antibodies (MA et al. 1995) have all been produced in leaves of tobacco following trafficking through the endoplasmic reticulum (ER) and Golgi complex. Human serum albumin has also been stably expressed in tobacco leaves and various tissues of potato including tubers (SIJMONS et al. 1990), although the precise folding and functionality of the protein was not established. In the above examples the recombinant proteins were either specifically targeted to, and detected in the apoplast, or presumed to locate there as a result of the default pathway of the plant endomembrane system. Deposition into the extracellular apoplast may contribute to the stability of foreign proteins by removing them from the more hydrolytic intracellular environment (FIREK et al. 1993).

Expression and accumulation of foreign proteins in seeds may be achieved through compartmentalization within various subcellular storage organelles. As a natural storage organ, seeds possess attributes such as high protein content and a low hydrolytic intracellular environment that make them particularly attractive as protein production vehicles. The human neuropeptide, leu-enkephalin (VAN-DEKERCKHOVE et al. 1989), and the leech anticoagulant protein, hirudin (PARMEN-TER et al. 1995), have both been produced in seeds of *Brassica napus* following targeting to protein bodies and oil bodies respectively. Proteins can also be secreted to the apoplast of seeds. However, the recovery of apoplastic proteins from seeds may be more difficult than from some of the vegetative organs mentioned above, owing to the dessicated state of seed tissue at maturity. On the other hand, this advanced state of dehydration also confers enhanced stability, allowing seeds to be stored for periods of several years without any appreciable degradation of proteins or loss of activity (e.g., see PEN et al. 1993). The greater flexibility resulting from the separation of protein production and purification represents a distinct advantage of seeds over most other organs for which more immediate processing is often required.

2.3 Expression Strategies

Choice of promoter, which mediates the timing, tissue-specificity, and level of transgene expression, is a key determinant of transgene product yields and recovery strategies (see review by CHINN and COMAI 1996). As shown in Table 2, many of the human (or other animal) proteins expressed in plants have used native or enhanced versions of the 35S promoter derived from the cauliflower mosaic virus to drive "constitutive" transgene expression, and it remains the most widely used promoter in plant biology for over-expression of plant proteins or inhibition via antisense strategies. The 35S promoter is active in most plant tissues (BENFEY et al. 1989; FANG et al. 1989) and especially in its modified forms (KAY et al. 1987; CARRINGTON and FREED 1990) can drive quite high levels of protein production. Although most of the human proteins produced using the 35S promoter (Table 2) showed accumulation levels below 0.1% of soluble protein, several transgene products (chitinase, antibodies) have been expressed at levels of 2%–5% of extractable protein. The 35S promoter is quite active during seed development and has been used in production systems targeting recovery of recombinant proteins from seed. However, the 35S promoter (and constitutive expression in general) has significant limitations when commercial bioproduction in nonseed tissues is the goal. Proteins that accumulate to high levels may negatively impact yield or overall health of the plant. High constitutive expression is sometimes associated with co-suppression or gene silencing (TAYLOR 1997) resulting in little or no transgene product accumulation. For proteins that are not highly stable, constitutive expression can lead to wasteful synthesis-degradation cycles and, of particular detriment for pharmaceutical application, contamination of the final product with inactive degradation products. In addition, the 35S is not highly active in many mature tissues (e.g., mature roots and fully expanded leaves) so that the full potential of biomass cannot be utilized. Use of inducible promoters or promoters that have a tight pattern of tissue- or organ-specificity avoids many of these limitations and appears to be the strategy of choice for most companies targeting plant-based production of high-value proteins.

CropTech scientists have developed a postharvest expression system that uses an inducible promoter termed the MeGA™ promoter (CRAMER and WEISSENBORN 1997). This promoter has been modified from a defense-related gene such that it is generally inactive during normal growth and development but shows rapid and strong gene activation in response to mechanical stress (wound-induction, or mechanical gene activation) or a variety of defense elicitors. Thus, the recombinant protein is not synthesized in tobacco leaves in the field (or greenhouse). Plants can be harvested and stored for weeks in a cold room. Recombinant protein production is then induced *de novo* in the laboratory or GMP facility and newly synthesized protein recovered 8–24h later. Because survival depends on both the speed and intensity with which a plant can activate its defenses, we find the MeGA™ promoter highly effective in driving high levels of inducible expression in all tissues of the plant including fully expanded leaves. The postharvest expression strategy has several advantages for pharmaceutical production. Biomass production is both temporally and spatially separated from recombinant product production minimizing the impact of (a) environmental factors on protein yield and quality and (b) possible deleterious effects of transgene expression or foreign protein accumulation on plant growth and development. All recovered protein is newly synthesized. In addition, the timing of protein extraction can be adjusted based on the stability of the particular gene product to optimize yield of fully active polypeptides. For products requiring activation of multiple genes (e.g., multiple subunits, or target proteins that require specialized protein-modifying enzymes), coinduction assures coordinated synthesis. In theory, the postharvest system could also permit further manipulation of the protein synthesis and processing machinery through addition of specific chemicals to the induction medium (e.g., inhibitors of key protein modification steps), although this could add significant expense to commercial scale bioproduction.

Bioproduction strategies involving developmentally defined- or virally vectored-expression (e.g., Biosource's Geneware system) are also designed to limit recombinant protein production to a discrete period. With the Geneware system, TMV-susceptible tobacco is field grown to an appropriate age, inoculated with genetically modified virus, and harvested 2–3 weeks later for recombinant protein extraction (GRILL 1997). Within this period, the virus titers reach high levels leading to significant transgene product accumulation. Using this system, Biosource scientists have attained very high product yield (recombinant protein representing greater than 10% of total soluble protein) and have progressed to the point of large scale field production and pilot plant extraction. Applied Phytologics utilizes a germination-specific promoter to direct transgene expression. Recombinant protein is produced under controlled conditions following imbibition and initiation of germination of transgenic seed, a production scheme analogous to barley malting. Expression strategies involving seed-based accumulation of recombinant proteins also take advantage of discrete bioproduction periods and separation of transgene activity from the bulk of plant growth. A large number of seed-specific promoters, often derived from genes encoding seed storage proteins, are available for both monocot and dicot plants. Depending on the recovery strategy (see below) and the characteristics of the protein product, promoters specific for embryo- versus endosperm-specific expression can be selected.

2.4 Posttranslational Processing

In comparison with industrial enzyme production, bioproduction of human proteins for pharmaceutical applications is particularly challenging due to the rigorous requirements with respect to purity, reproducibility, efficacy, and biocomparability. Many of the proteins with greatest promise as therapeutics require complex posttranslational modifications and/or assembly. The striking fidelity with which plants appear to recognize and correctly act upon most of the processing signals encrypted within mammalian polypeptides indicates a high degree of conservation in protein processing machinery between plants and animals. Conserved processes include endomembrane targeting, signal peptide cleavage, protein folding and oligomerization, disulfide bond formation (although precise cysteine-cysteine bonding patterns have not been directly determined), asparagine-linked glycosylation, selective retention in the ER and Golgi, and C-terminal isoprenylation. We have also noted internal proteolytic processing events in several human proteins expressed in tobacco that appear to mimic processing that occurs in mammalian cells although the precise termini of the products have not yet been determined (Oishi et al. unpublished data).

However, clear differences in protein processing, most notably in glycoprotein processing, do exist between plants and animals. The glycan moiety of mammalian glycoproteins functions in protein folding and assembly, subcellular targeting, cell- or tissue-specific delivery within the body, protein half-life, and clearance from the bloodstream (VARKI 1993). Thus, changes in glycan composition or arrangement are likely to affect activity or pharmacokinetics (JENKINS et al. 1996; LUI 1992). Plant N-linked glycans do not contain terminal sialic acid residues or mannose-6-phosphate and contain other sugars or sugar linkages not found in mammalian glycoproteins. The same amino acid signature (N-X-S/T) is recognized within the ER for addition of the high-mannose form glycan complex (identical in plant and animals). However, plants process those N-linked glycans to distinct complex forms as the glycoprotein progresses through the Golgi. The sialic acid is present as the terminal sugar on many serum glycoproteins and appears to function in serum longevity and rates of clearance for some serum proteins (GRINNELL et al. 1991). Incorporation of this charged sugar residue into protein glycans has not been demonstrated in plants (FAYE et al. 1989). In addition, plants do not phosphorylate high-mannose glycans – in mammals, the mannose-6-phosphate serves as a signal to target soluble glycoproteins to lysosomes. Finally, many complex plant glycans contain either fucose or xylose residues with linkages that do not occur in humans. Plant-synthesized glycoproteins displaying these sugar linkages appear highly im- munogenic when injected into mammals (CRISPEELS and FAYE 1996). Interestingly, an Arabidopsis mutant defective in N-acetylglucosaminyl-transferase-I has been identified in which all N-linked glycans are in the high-mannose form (VON SCHAEWEN et al. 1993). This report suggests that processing of glycans to complex forms is not critical for plant viability or development (in contrast to animals). Thus, plants can be altered to produce nonimmunogenic glycans. Variations in glycan composition is not unique to plant-based recombinant systems – yeast, baculovirus/insect cell, transgenic animal milk and even mammalian cell cultures often generate glycans that are heterogeneous or differ significantly from the native conformation for particular human proteins (reviewed in JENKINS et al. 1996). It is clear that additional research is required for effective bioproduction of human gly- coproteins in plants (discussed further for lysosomal proteins in Sect. 3). Genetic engineering strategies to modify the glycan-processing machinery of plants or *in vitro* enzymatic modification of the purified recombinant protein should enable com- mercialization of plant-synthesized glycoproteins for pharmaceutical applications.

Because plants are relatively easy to genetically engineer, genetic strategies to specifically alter protein processing by either antisense to block endogenous enzymes or addition of genes encoding novel processing activities are highly feasible. The recent cloning of plant genes encoding enzymes involved in Golgi- localized glycan processing opens up opportunities to modify the complex glycans produced in plants. Processes other than glycosylation can also be modified. We are interested in testing whether plants can be engineered to produce the complex serum proteinases involved in the coagulation-anticoagulation cascade (CRAMER et al. 1996a; WEISSENBORN et al. 1995). Plants are unlikely to perform the highly specialized γ-carboxylation of the amino-terminal glutamates required for bioactivity of several of these enzymes (protein C, thrombin, clotting factors VII, IX and X). We are currently introducing a human cDNA for the vitamin K dependent γ-carboxylase to perform the necessary modifications for this class of proteinase into tobacco (Cramer, Grabau, et al. unpublished data). While these experiments are in very early stages, the concept of engineering elite plant lines for specialized protein processing for pharmaceutical bioproduction seems highly feasible.

2.5 Recovery Strategies

To capitalize on the advantages of plant-based systems in upstream production, it is necessary that downstream purification of the recombinant product be accomplished economically. Complex and inefficient purification schemes can contribute significantly to overall costs and result in lower yields so that commercial production is no longer viable. In some cases, such as in the production of industrial enzymes, downstream costs can be reduced or even eliminated when a high degree of product purity is not required. A good example of this is the production of phytase in seeds. The enzyme phytase may be used to enhance the nutritional quality of seed meal by breaking down the phytase present in the meal and thereby increasing the availability of phosphate to monogastric animals. This may be conveniently achieved through expressing the phytase enzymes in seeds and adding milled transgenic seed to a standard feed meal preparation (PEN et al. 1993; VERWOERD and PEN 1996). Unfortunately, this strategy is not applicable to many proteins, particularly pharmaceutical proteins, that require rigorous purification to near-homogeneity. For these products simple and efficient methods of downstream purification must be developed.

2.5.1 Affinity Tag-Based Purification

One approach to the purification of recombinant proteins is through the use of affinity tags. This can be accomplished through the creation of a fusion between the protein of interest and another protein or peptide that exhibits affinity for a specific ligand. The fusion protein is then recovered by binding to the ligand immobilized onto a support matrix. The high selectivity possible with affinity separation often enables a substantial degree of purification to be achieved in a single step. A number of these affinity tags have been developed for use in microbial systems. Different types of ligand pairs have been exploited for this purpose including maltose binding protein–amylose, histidine residues–metal ions, and protein A–IgG. A similar approach may be useful for the purification of recombinant proteins synthesized in plants. The efficacy of this method in plants has been demonstrated in a small scale purification of a human glucocerebrosidase-FLAG epitope fusion produced in tobacco (CRAMER et al. 1996b). Here, the fusion protein was recovered using an anti-FLAG antibody affinity matrix and used for biochemical studies on activity and posttranslational modifications. However, because the long-term application is as a replacement enzyme therapeutic for Gaucher patients, the presence of the "nonhuman" residues is undesirable and is not used for scale-up. For some proteins and production strategies, the affinity tag can be proteolytically removed from the fusion protein following purification. However, as with any strategy involving cleavage of fusion proteins, the additional steps required to remove the tag contribute to the downstream purification costs, and there is the potential that the tag could alter folding or processing of the recombinant protein.

2.5.2 Compartmentalization

Another means of simplifying the purification of recombinant proteins is through compartmentalization. This can be achieved using either signal peptides or whole protein fusions to target the protein to a specific cellular location. In this case, purification of the desired protein is facilitated by virtue of its physical separation from other proteins in the cell. Subcellular fractionation is then used to obtain an enriched fraction containing the recombinant protein. A variety of forms of compartmentalization have been recruited for the production of foreign proteins in plants. These include expression on viral particles, extracellular secretion, and targeting to intracellular storage organelles. As noted above, the posttranslational modifications required to produce a functional protein necessarily constrain where that protein can be expressed, as these reactions are, to a large extent, localized to specific subcellular compartments.

A number of plant viruses have been used for the transient expression of foreign proteins in plants (DE ZOETEN et al. 1989; GRILL 1993; USHA et al. 1993). To aid in purification, recombinant proteins may be engineered as fusions to viral coat proteins and then separated as mature virus particles. This strategy has recently been used in the production of malarial antigens in tobacco with tobacco mosaic virus (TMV) (TURPEN et al. 1995). Nontransgenic plants were inoculated with infectious RNA transcribed from cDNA encoding the genetically engineered fusion protein. Mature viral particles carrying the fusion were later recovered from leaf extracts through differential centrifugation and precipitation. While in this example the purified viral particles were intended for use as a malaria vaccine, it should also be possible to further purify recombinant proteins with this approach by introducing a protease cleavage site into the fusion protein. One possible limitation to this approach may be the size of the foreign protein, as larger proteins may impair viral coat assembly.

Secretion into the extracellular media or periplasmic space has proven to be extremely useful for production and purification of foreign proteins in many yeast and bacterial systems. In addition to providing an enriched fraction of the recombinant product, secretion has also been found to enhance protein stability and facilitate proper folding. Another attractive feature of this approach is that the signal peptide is removed from the recombinant protein in the course of normal processing enabling an authentic protein to be obtained without introducing additional proteolytic digestion steps. In plant cells, secreted proteins are deposited into the apoplastic space. The native signal peptide as well as a signal sequence from the tobacco pathogenesis-related protein, PR-S, have been used to success-fully direct secretion of human serum albumin in potato (SIJMONS et al. 1990). Similarly, the potato proteinase inhibitor II protein signal peptide (HERBERS et al. 1995) has been used to secrete xylanase into the apoplastic space of tobacco plants. While considerable enrichment of the recombinant protein can be achieved with this approach, methods of efficiently recovering proteins from the apoplastic fluid have yet to be developed.

With the appropriate signals or fusions it is also possible to target proteins to the lumen of the ER or vacuole. The human neuropeptide leu-enkephalin has been expressed in seeds of *Arabidopsis thaliana* and *Brassica napus* as an internal fusion between the N- and C-terminal ends of the Arabidopsis 2S albumin protein (VANDEKERCKHOVE et al. 1989). The fusion protein was subsequently found to accumulate stably within the protein bodies of these seeds. Purification was ac-complished through an initial fractionation in low salt to obtain albumin proteins followed by two proteolytic digestion steps and HPLC separation. One drawback of this strategy is the complexity of the proteolytic cleavage, particularly since carboxypeptidase was required to remove the C-terminal portion of the albumin protein. A failure to precisely control this reaction would result in significant product heterogeneity. It is also possible that folding constraints for protein body packaging might impose limitations on the size of the foreign protein that could be produced as an internal fusion.

Seed oil bodies represent another subcellular compartment available for tar-geting of recombinant proteins. Localization to oil bodies is achieved through creating a fusion between the desired recombinant protein and oleosin, a protein specifically targeted to these organelles. As described below, oil bodies offer some unique advantages and opportunities for expression and purification of foreign proteins.

2.5.3 Seed Oil Bodies as Purification Tools

Oil bodies are natural subcellular organelles found in all oilseeds where they form the storage site for the primary energy reserve in these seeds, triacylglycerides (TAGs). They are comprised of TAGs surrounded by a half-unit phospholipid membrane into which is embedded a unique type of protein known as oleosin. Oleosins accumulate to high levels in oil seeds comprising between 2% and 10% of the total seed protein in different species. It is believed that the primary function of oleosins is to prevent the coalescence of oil bodies during seed desiccation. In so doing, a larger surface area is available for lipolytic enzymes enabling the rapid mobilization of TAG reserves upon seed germination. Although the precise mechanism of oleosin targeting is not fully understood, it is known that they are synthesized on the ER and that a motif in the central domain is crucial for their subsequent localization to oil bodies (VAN ROOIJEN and MOLONEY 1995a; ABELL et al. 1997). The oleosin protein appears to consist of three distinct domains. The N- and C-terminal domains are amphipathic and proteolytic digestion studies strongly suggest that they reside on the outer surface of the oil body (ABELL et al.

1997; HILLS et al. 1993; TZEN and HUANG 1992). The central domain is comprised
largely of hydrophobic amino acid residues, and is believed to adopt a hairpin
conformation anchoring the protein firmly within the TAG core of the oil body.
Comparison of oleosin sequences from different species reveals that the central
domain is highly conserved while the N- and C-termini exhibit considerable sequence variation.

Several features of seed oil bodies lend themselves to the production of foreign
proteins. Oleosins tolerate fusion of foreign proteins to either the N- or C-terminal
ends without apparent loss of oil body targeting efficiency (MOLONEY and VAN
ROOIJEN 1996). Oleosin fusions have been created with a number of different
proteins varying in molecular weight from approximately 7–55kDa, all of which are
stably accumulated on the surface of oil bodies. In the case of the reporter enzyme
β-glucuronidase, it was further shown that enzymatic activity was retained with the
oleosin fusion-oil body complex. The oil bodies, together with their complement of
oleosin proteins, are remarkably stable both within the seed and following their
release by aqueous extraction (VAN ROOIJEN and MOLONEY 1995b) Within the seed
the proteins remain undegraded for years without the requirement for elaborate
storage conditions. Following their release into aqueous solution, oil bodies are
extremely resistant to mechanical disruption and are stable over a wide range in pH
and temperature (KÜHNEL et al. 1996; VAN ROOIJEN and MOLONEY 1995b). Finally,
the lower density of oil bodies allows them to be separated from soluble contaminants by flotation centrifugation, enabling simple and rapid purification of recombinant proteins targeted to the oil body surface. Digestion with a site-specific
endoproteinase to cleave the oleosin fusion protein, and centrifugation to remove
the oil bodies, results in the recovery of a highly enriched fraction of the desired
recombinant protein within the aqueous phase. The naturally low hydrolytic environment within the seed, coupled with the rapid removal of soluble-protein
contaminants, ensures that little or no degradation of the oil body-associated
proteins occurs during processing. As described in Sect 3.2, the unique properties of
oleosins and oil bodies have been exploited by Sem BioSys in the development of a
novel plant-based protein production and purification system.

3 Examples of Plant-synthesized Protein Therapeutics: Linking Upstream and Downstream Strategies

In order to "reduce to practice" many of the considerations and strategies described above, two very different examples of plant-based bioproduction of recombinant proteins of commercial value are described below. These examples not
only demonstrate the diversity of expression and purification strategies available
through plants, but also highlight the constraints on bioproduction strategies imposed by the particular protein target. In both cases, the overall bioproduction
strategy has been strongly influenced by commercial and regulatory considerations.

3.1 Production of Human Lysosomal Enzymes
in *Nicotiana tabacum*

Cost-effective production of recombinant human proteins for replacement enzyme therapies is likely to have a huge impact on the care and treatment of patients with specific metabolic or genetic disorders. The lysosomal storage disorders represent a large class of these genetic diseases for which the molecular basis of disease has been determined and cDNAs encoding the required enzymes have been cloned (NEUFELD 1991). Lysosomes, the animal organelle responsible for the regulated intracellular degradation of macromolecules, contain multiple hydrolases including proteases, nucleases, glycosidases, lipases, phosphatases, phospholipases, and sulfatases (DARNELL et al. 1986). Deficiency in specific lysosomal hydrolases can lead to toxic accumulation of the undegraded substrate and a variety of clinical manifestations. Tay-Sachs disease is perhaps the most familiar lysosomal storage disease, involving deficiencies in α-hexosaminidase that lead to accumulation of ganglioside G_{M2} in the membranes of brain cells (NEUFELD 1991). The mucopolysaccharidoses (MPSs) are a group of lysosomal storage diseases caused by deficiencies of one or more of the ten lysosomal enzymes required for the degradation of sulfated glycosaminoglycans (reviewed in NEUFELD and MUENZER 1995). Lysosomal accumulation of undegraded glycans leads to the malfunction of affected cells/organs which compromises the growth and development of the individual and may, in severe cases, lead to premature death. Replacement enzyme therapy appears promising based on human cell- and animal models, but drug development is hampered by the small patient pool and limitation in current technologies for cost-effective bioproduction. The industry paradigm for human replacement enzyme therapy is the glycoprotein product Ceredase (Genzyme, Cambridge, MA) for the treatment of Gaucher disease. This lysosomal storage disorder affects 10,000–20,000 individuals in the United States (NIH TECHNOLOGY ASSESSMENT PANEL ON GAUCHER DISEASE 1996) and is caused by defects in glucocerebrosidase, an acid β-glucosidase required for complex lipid degradation. Routine administration (generally every 2 weeks) of placental-derived enzyme has revolutionized the treatment of the disease and the quality of life of Gaucher patients. However, the high drug costs associated with purification of glucocerebrosidase from human placentae or, more recently, with bioproduction of recombinant enzyme in Chinese hamster ovary (CHO) cells, make it one of the world's most expensive drugs. Although the production of lysosomal enzymes in plants is challenging (CRAMER et al. 1996b), CropTech has selected several lysosomal enzymes among its initial targets for bioproduction based on (a) the ability of plants to address critical cost, safety and supply issues for replacement enzymes, (b) the extreme medical need, and (c) the potential for Orphan Drug status to facilitate progress toward clinical trials and commercialization.

The first lysosomal enzyme produced in transgenic plants was glucocerebrosidase (EC3.2.1.45) as a potential alternative replacement therapy for Gaucher disease (CRAMER et al. 1996a,b). Placental glucocerebrosidase that has been enzymatically modified to generate mannose-terminated glycans is highly effective in reversing the symptoms of the disease (BRADY et al. 1974). For commercial production of glucocerebrosidase (Ceredase, Genzyme), it requires between 400 and 2000 placentae to supply a standard dose – a major factor in the extreme cost to patients ($100,000–400,000 annually; NIH TECHNOLOGY ASSESSMENT PANEL ON GAUCHER DISEASE 1996). A CHO-synthesized recombinant form (Cerezyme, Genzyme) has been approved by the FDA, but no significant reduction in cost is anticipated. As a consequence, the success of treatment of Gaucher's disease in the United States remains limited by the cost and supply of the drug. For these reasons, it is a promising candidate for production in a plant-based system. The successful synthesis of enzymatically active human glucocerebrosidase in transgenic tobacco plants was previously described (CRAMER et al. 1996a,b). Briefly, the entire coding region for human glucocerebrosidase was modified to encode a FLAG (International Biotechnologies) epitope tag for subsequent detection and purification, fused to the inducible MeGA^{TM} promoter (CropTech Corporation), and introduced into tobacco. Of 49 transgenic plants analyzed, five plants had expression levels between 1%–10% total soluble leaf protein after 12h of postharvest induction. Based on biochemical kinetic studies, affinity binding characteristics, and conduritol B inhibition, the tobacco-synthesized enzyme appears fully active and comparable to human- or CHO-derived glucocerebrosidase. These studies form the basis of current and future efforts in the commercial production of glucocerebrosidase in transgenic plants for Gaucher enzyme replacement therapy. Studies to (a) identify high-expressing transgenic lines containing new glucocerebrosidase constructs that lack the nonhuman FLAG epitope and (b) test strategies to address glycan modification are now underway prior to scale-up of glucocerebrosidase production and purification technologies.

CropTech researchers have also synthesized a second lysosomal enzyme, α-L-iduronidase (IDUA, EC3.2.1.76), in transgenic tobacco (Jenkins, Weissenborn, Bennett, and Oishi, unpublished results). IDUA is a potential replacement therapeutic for Hurler syndrome and Hurler/Scheie syndrome, the most common MPS representing 1/100,000–1/150,000 births. Although the concept of enzyme replacement for Hurler syndrome was first investigated in the 1970s (DIFERRANTE and NICHOLDS 1974; NEUFELD and MUENZER 1989), the development of IDUA as a drug has not progressed rapidly because of the lack of an effective production system. Recombinant enzyme sufficient for initial testing in Hurler-canine and feline models has been produced using a CHO-based production system (KAKKIS et al. 1994), but progress toward human trials is limited by protein availability. As a consequence, the successful plant-based production of IDUA has the potential to directly impact the speed of development of IDUA as a enzyme replacement therapy for Hurler and Hurler/Scheie syndromes. In humans, the lysosomal IDUA from liver is a soluble glycoprotein of 60–82kDa reflecting heterogeneity in the glycan composition. There are six potential N-linked glycosylation sites, some of which are modified to mannose-6-phosphate forms (generally sites 3 and 6) or to complex glycan forms. At all sites there is a high degree of microheterogeneity in glycans (ZHAO et al. 1996). The sequence of the complete cDNA for human IDUA has been reported (MOSKOWITZ et al. 1992; SCOTT et al. 1991) and encodes a protein of 653 amino acids (pre-IDUA) with a signal peptidase cleavage site at amino acid 27. The cDNA for IDUA has been expressed in Cos-1 and CHO cells (KAKKIS et al. 1994; SCOTT et al. 1991) and recombinant IDUA has been purified and shown to be biologically active in dogs deficient for this enzyme (KAKKIS et al. 1996).

As an initial test of the feasibility of commercial production of human IDUA in plants, researchers at CropTech Corporation engineered tobacco plants for postharvest production of human IDUA. The coding region from the human IDUA cDNA (KAKKIS et al. 1994) was placed downstream of the inducible MeGA^{TM} promoter and inserted into a plant transformation vector for *Agrobacterium*-mediated transformation (HORSCH et al. 1985). Transgenic tobacco plants containing one to three copies of the MeGA promoter-IDUA construct were tested for IDUA expression. In noninduced maturing transgenic leaf material, no IDUA mRNA was detected. Following 8–24h of induction, abundant IDUA transcripts were detected as well as novel proteins that cross-reacted with anti-IDUA antisera. Tobacco contains no endogenous IDUA activity (in contrast to the acid β-glucosidases which are quite abundant in tobacco). Analyses of IDUA recovered from induced tobacco leaf material indicated that the IDUA protein is enzymatically active. Because human IDUA requires glycosylation for activity, detection of enzyme activity in tobacco samples (as well as its electrophoretic mobility and ConA binding) indicates that the human IDUA signal peptide correctly targets the protein to the plant endomembrane system for glycosylation. The majority of the IDUA appears to be secreted, the default pathway for the plant endomembrane system (DENECKE et al. 1990). Although IDUA yields from the first IDUA-transgenic plants analyzed are lower than those seen for glucocerebrosidase-expressing plants, demonstrations of enzymatic activity of the tobacco-synthesized IDUA glycoprotein and development of novel IDUA recovery methods strongly support the use of transgenic tobacco for human IDUA production.

Both glucocerebrosidase and IDUA are glycoproteins and thus pose a particular challenge for production in plants as well as other recombinant expression systems (JENKINS et al. 1996). For soluble lysosomal enzymes such as IDUA, the signal for lysosomal sorting is the mannose-6 phosphate residues present in their N-linked glycans. Mannose-6-phosphate receptors are present on the plasmamembrane as well as lysosomal membranes of many mammalian cell types and thus direct uptake and lysosomal delivery of exogenously supplied IDUA (KAKKIS et al. 1996). Plants do not phosphorylate their glycans and glycan-based signals do not appear to function in vacuolar targeting (FAYE et al. 1989; CHRISPEELS and FAYE 1996). It is likely that some, if not all, of the glycans on tobacco-synthesized IDUA are in the complex form and thus likely to be immunogenic (see Sect. 2.4) and ineffective in directing the required cell-specificity for uptake and lysosomal delivery. Engineering plants to synthesize mannose-6-phosphate-modified glycans is currently not feasible — the two required enzymes have not been well characterized. However, alternative strategies that address both the delivery and immunogenicity are suggested by the currently effective lysosomal replacement therapeutic, Ceredase. Glucocerebrosidase is a membrane-associated protein that is targeted to lysosomes by a mannose-6-phosphate-independent route. The N-linked glycans present on the placental enzyme are bianteniary structures having terminal sialic acid residues. In order to direct effective delivery to lysosomes of the affected cells in Gaucher patients (primarily cells of the macrophage/monocyte lineage), sequential enzymatic digestion is used to remove the terminal sugars and expose the mannose core (BARTON et al. 1991). This mannose-terminated form is targeted to the correct cell and organellar location to effect glucoceremide degradation and symptom reduction (GRABOWSKI et al. 1995; BARTON et al. 1991). Complex plant glycans are naturally mannose terminated (CHRISPEELS and FAYE 1996). Enzymatic removal of the immunogenic fucose and xylose residues should yield glycans of similar pharmacokinetics as Ceredase.

3.2 Production of Hirudin in *Brassica napus*

To evaluate the potential of Sem BioSys' oleosin partitioning technology, the model therapeutic protein hirudin was selected. Hirudin is a naturally occurring anticoagulant protein produced in the salivary glands of medicinal leeches (*Hirudo medicinalis*) and secreted to facilitate feeding. Since its discovery almost 50 years ago, it has been extensively studied. Hirudin possesses a number of desirable properties which advocate its use as a therapeutic pharmaceutical. It is an extremely specific and potent inhibitor of thrombin, the last enzyme in the blood coagulation cascade, having a KI of 21fM (BRAUN et al. 1988). It is also rapidly cleared from the body, exhibits low toxicity (500,000U/kg body weight in rats) (MARKWARDT et al. 1982) and, probably as a consequence of the coevolution of leeches and mammals, has relatively low immunogenicity (KLOCKING 1991). The protein has also been well characterized with respect to its structure and mechanism of binding to thrombin (RYDEL et al. 1990). A small number of closely related isoforms of hirudin have been isolated all of which show strict conservation for six cysteine residues (STONE and MARAGANORE 1993). These residues participate in the formation of three disulfide bridges whose precise pairing is necessary for protein activity (CHATRENET and CHANG 1992, 1993). Although the native protein is sulfated at the Tyr-63 position, recombinant nonsulfated hirudin exhibits significant activity (STONE and MARAGANORE 1993). It folds spontaneously *in vitro* and functional hirudin has been produced previously in both bacterial (HARVEY et al. 1986; BENDER et al. 1990) and yeast (LOISON et al. 1988; LEHMAN et al. 1993) systems. However, the quantities of hirudin required, were it to fully replace presently used anticoagulants such as heparin, are estimated to be on the order of hundreds to thousands of kilograms of protein annually. For this reason, hirudin is an excellent candidate for production with a high capacity plant-based system.

The common oilseed rape species, *Brassica napus*, was selected as the vehicle for production of seed-derived hirudin. After tobacco, the *Brassica* species are among those most easily transformed with *Agrobacterium*. Cells in the ends of cotyledonary petioles cut from young seedlings are readily infected with the bacterium. Formation of callus, regeneration to plants, and selection of transformants are all very efficient. In *B. napus*, transformation efficiencies approaching 55% of the original explants can be obtained. The time-line for development of a transgenic plant is also relatively short, in the range of approximately 4–6 months from transformation to collection of first generation transformed seed. Another attractive feature is the availability of a haploid production system from microspore-derived embryos, facilitating the creation of homozygous lines. As an oilseed crop, considerable biomass is concentrated within the seed. Seed production in *B. napus* is between 1 and 2 tons per hectare at a cost of approximately (United States) $300/ton. Protein content in these seeds represents in excess of 20% of the total seed weight, approximately 9% of which is oleosin.

The production and analysis of transgenic plants expressing an oleosin-hirudin fusion has been reported previously (PARMENTER et al. 1995). Briefly, a synthetic sequence encoding the hirudin variant 2 (HV2) isoform was fused to the 3' end of an Arabidopsis 19kDa oleosin gene with the two coding regions separated by a sequence encoding the four amino acid recognition site for the protease, factor Xa. Following *Agrobacterium*-mediated transformation, putative transgenics were selected and expression of the oleosin-hirudin fusion confirmed by northern analysis. Immunoblotting with anti-hirudin antibodies demonstrated that the oleosin-hirudin fusion protein was correctly targeted and accumulated on oil bodies of transgenic seed. Oil bodies were separated and washed to remove contaminating soluble proteins through flotation-centrifugation. After digestion with factor Xa and a final round of flotation-centrifugation to remove oil bodies, hirudin was recovered in the aqueous fraction. Formation of a functional protein was confirmed by an *in vitro* thrombin inhibition assay. Comparison of protein contents in whole seed extracts and in the soluble fraction obtained after flotation-centrifugation indicated that the majority of seed protein had been removed. The enrichment obtained with this procedure demonstrates the utility of oil body compartmentalization for purification of recombinant proteins. Further purification of the recombinant hirudin to near-homogeneity was achieved through anion exchange and reverse phase chromatography. Values obtained for the specific activity of *B. napus*-derived hirudin are equivalent to those reported for recombinant hirudin produced in yeast systems (LOISON et al. 1988).

3.2.1 Prospects of Oleosin-Partitioning Technology

The potential for commercial application of oleosin partitioning technology can be evaluated by examining the system with reference to certain key production parameters namely, production capacity, authenticity/functionality of product, downstream purification costs, and process scaleability. We have estimated the level of expression of the oleosin-hirudin fusion protein in our transgenic seed to be approximately 10% of that of the endogenous oleosin (PARMENTER et al. 1995). Based on this estimate, hirudin would represent approximately 0.3% of the total seed protein. While encouraging, this level is still somewhat lower than would be desired for a commercial production system. To increase expression levels, we are currently testing a number of strong seed-specific promoters other than oleosin in our fusion constructs. An increase in the expression of recombinant protein to the relatively modest level of 1% of seed protein would result in a system capacity of approximately 2kg of product per ton of seed. When coupled with low production costs and cost-effective purification, this level is within the range required for commercial viability.

The downstream purification of proteins synthesized as oleosin fusions is greatly simplified by the oil body separation process. However, in order for this process to be cost-effective, the fusion protein cleavage step must be both efficient and economical. While useful for demonstration purposes, the factor Xa used in our initial hirudin studies fails to meet these requirements. The enzyme is expensive, gave incomplete cleavage, and represented a contaminant which had to be removed in subsequent purification steps. To address this problem we are presently expressing proteases as oleosin fusions immobilized on the surface of oil bodies. This will enable both economical production of the protease and easy removal following fusion protein cleavage through the existing oil body separation process. A number of suitable candidate proteases have been identified and are currently being tested.

The importance of process scale-up in determining economic feasibility is often overlooked in the initial research and development phase of a new technology. Procedures that work well for typical laboratory scale experiments cannot always be directly scaled up or easily adapted to existing industrial processes. In the case of oleosin partitioning technology, we have developed and tested methods using industrial equipment for the large scale preparation of oil bodies. The results from these tests indicate that the process can be easily scaled up to meet commercial production requirements.

The recovery of active products such as hirudin and β-glucuronidase from oleosin fusions demonstrates that functional proteins can be produced using oleosin partitioning technology. However, the fact that oleosins are not exposed to the lumen of the ER either during synthesis or subsequent targeting to oil bodies, limits the range of different products that can be produced through oleosin partitioning. Proteins requiring glycosylation or other forms of posttranslational modification associated with passage through the endomembrane system would not be properly processed as oleosin fusions. Nevertheless, a large number of proteins are still amenable to production using this technology. In addition to therapeutic proteins, the list includes many food and industrial enzymes. Some of these products are presently under development. Additionally, the ability to produce functional proteins on the surface of oil bodies offers exciting new possibilities for the production of immobilized protein matrices (KÜHNEL et al. 1996). With continued development in each of the areas mentioned above, prospects for the successful commercialization of oleosin partitioning technology appear very promising.

4 Summary

We have described two very different and innovative plant-based production systems – postharvest production and recovery of recombinant product from tobacco leaves using an inducible promoter and oleosin-mediated recovery of recombinant product from oilseeds using a seed-specific promoter. Both base technologies are broadly applicable to numerous classes of pharmaceutical and industrial proteins. As with any emerging technology, the key to success may lie in identifying those products and applications that would most benefit from the unique advantages offered by each system. The postharvest tobacco leaf system appears effective for proteins requiring complex posttranslational processing and endomembrane targeting. Because of the remarkable fecundity and biomass production capacity of tobacco, biomass scale-up is very rapid and production costs are low. Clearly the development of equally cost-effective extraction and purification technologies will be critical for full realization of the commercial opportunities afforded by transgenic plant-based bioproduction. The recovery of protein from tobacco leaves or oleosin-partitioned proteins by oil-body separations represent significant breakthroughs for cost-effective commercialization strategies. Additional low-cost, high-affinity separation technologies need to be developed for effective scale-up purification of plant-synthesized recombinant proteins. Clearly successful commercialization of plant-synthesized biopharmaceuticals must effectively link upstream strategies involving gene and protein design with downstream strategies for reproducible GMP-level recovery of bioactive recombinant protein. Both the tobacco and oilseed systems are uniquely designed to address issues of biomass storage, product recovery, quality assurance, and regulatory scrutiny in addition to issues of transgene expression and protein processing.

References

Abell BM, Holbrook LA, Abenes M, Murphy DJ, Hills MJ, Moloney MM (1997) Role of the proline knot motif in oleosin endoplasmic reticulum topology and oil body targeting. Plant Cell 9:1481–1493

Barton NW, Brady RO, Dambrosia JM, Brisaceglie AM, Doppelt SH, Hill SC, Mankin HJ, Murray GJ, Parker RI, Argoff CE, Grewal RP, Yu K-T (1991) Replacement therapy of inherited enzyme deficiency: macrophage-targeted glucocerebrosidase for Gaucher's disease. N Engl J Med 324:1464

Bender E, Vogel R, Koller K, Engels J (1990) Synthesis and secretion of hirudin by *Streptomyces lividans*. Appl Microbiol Biotech 34:203–207

Benfey PN, Ling R, Chua N-H (1989) The CaMV 35 S enhancer contains at least two domains which can confer different developmental and tissue-specific expression patterns. EMBO J 8:2195–2202

Bosch D, Smal J, Krebber E (1994) A trout growth hormone is expressed, correctly folded and partially glycosylated in the leaves but not the seed of transgenic plants. Transgenic Res 3:304–310

Brady RO, Pentchev PG, Gal AE, Hibbert SR, Dekaban AS (1974) Replacement therapy for inherited enzyme deficiency. Use of purified glucocerebrosidase in Gaucher's disease. N Engl J Med 291:989–993

Braun PJ, Dennis S, Hofsteenge J, Stone SR (1988) Use of site-directed mutagenesis to investigate the basis for the specificity of hirudin. Biochemistry 27:6517–6522

Carrington JC, Freed DD (1990) Cap-independent enhancement of translation by a plant potyvirus 5' nontranslated region. J Virol 64:1590–1597

Chatrenet B, Chang J (1992) The folding of hirudin adopts a mechanism of trial and error. J Biol Chem 267:3038–3043

Chatrenet B, Chang J (1993) The disulfide folding pathway of hirudin elucidated by stop go folding experiments. J Biol Chem 268:20988–20996

Chinn AM, Comai L (1996) Gene transcription. In: Owen MRL, Pen J (eds) Transgenic plants: a production system for industrial and pharmaceutical proteins. Wiley, Chichester, pp 27–48

Chong DKX, Roberts W, Arakawa T, Illes K, Bagi G, Slattery CW, Landridge WHR (1997) Expression of the human milk protein b-casein in transgenic potato plants. Transgenic Res 6:289–296

Chrispeels MJ, Faye L (1996) The production of recombinant glycoproteins with defined non-immunogenic glycans. In: Owen MRL, Pen J (eds) Transgenic plants: a production system for industrial and pharmaceutical proteins. Wiley, Chichester, pp 99–114

Cramer CL, Weissenborn DL, Oishi KK, Grabau EA, Bennett S, Ponce E, Grabowski GA, Radin DN (1996a) Bioproduction of human enzymes in transgenic tobacco. In: Collins GB, Shepherd RJ (eds) Engineering Plants for Commercial Products and Applications. New York Academy of Sciences, New York, pp 62–71

Cramer CL, Weissenborn DL, Oishi KK, Radin DN (1996b) High-level production of enzymatically active human lysosomal proteins in transgenic plants. In: Owen MRL, Pen J (eds) Transgenic plants: a production system for industrial and pharmaceutical proteins. Wiley, Chichester, pp 299–310

Cramer CL, Weissenborn DL (1997) HMG2 promoter expression system and post-harvest production of gene products in plants and plant cell cultures. US. Patent 5,670,349

Darnell J, Lodish H, Baltimore D (1986) Molecular cell biology. Scientific American, New York

Denecke J, Botterman J, Deblaere R (1990) Protein secretion in plant cells can occur via a default pathway. Plant Cell 2:51–59

de Zoeten GA, Penswick JR, Horisberger MA, Ahl P, Schultze M, Hohn T (1989) The expression, localization and effect of a human interferon in plants. Virology 172:213–222

Dieryck W, Pagnier J, Poyart C, Marden MC, Gruber V, Bournat P, Baudino S, Merot B (1997) Human haemoglobin from transgenic tobacco. Nature 385:29–30

DiFerrante NM, Nicholds BL Jr (1974) Comments on the plasma treatment of the mucopolysaccharidoses. Birth Defects 10:234–238

Fang R-X, Nagy F, Sivasubramanian S, Chua N-H (1989) Multiple cis-regulatory elements for maximal expression of the cauliflower mosaic virus 35S promoter in transgenic plants. Plant Cell 1:141–150

Faye L, Johnson KD, Strum A, Chrispeels MJ (1989) Structure, biosynthesis, and function of asparagine-linked glycans on plant glycoproteins. Plant Physiol 75:309–314

Firek S, Draper J, Owen MRL, Gandecha A, Cockburn B, Whitelam GC (1993) Secretion of a functional single-chain Fv protein in transgenic tobacco plants and cell suspension cultures. Plant Mol Biol 23:861–870

Ganz PR, Dudani AK, Tackaberry ES, Sardana R, Sauder C, Cheng X, Altosaar I (1996) Expression of human blood proteins in transgenic plants: the cyokine GM-CSF as a model protein. In: Owen MRL, Pen J (eds) Transgenic plants: a production system for industrial and pharmaceutical proteins. Wiley, Chichester, pp 281–297

Grabowski GA, Barton NW, Pastores G, Drambrosia JM, Banerjee TK, McKee MA, Parker C, Schiffmann R, Hill SC, Brady RO (1995) Enzymes therapy in type 1 Gaucher disease: comparative efficacy of mannose-terminated glucocerebrosidase from natural and recombinant sources. Ann Intern Med 122:33–39

Grill LK (1993) Tobacco mosaic virus as a gene expression vector. Agro-Food-Industry Hi-Tech November/December: 20–23

Grill LK (1997) Viral-vectored, large scale production of drugs and pharmaceuticals in plants. Presentation at IBCs 3rd Annual International Symposim on Producing the Next Generation of Therapeutics: Exploiting Transgenic Technologies, West Palm Beach, 5–6 Feb

Grinnell BW, Walls JD, Gerlitz B (1991) Glycosylation of human protein C affects its secretion, processing, functional activities and activation by thrombin. J Biol Chem 226:9778–9785

Haq TA, Mason HS, Clements JD, Arntzen CJ (1995) Oral immunization with a recombinant bacterial antigen produced in transgenic plants. Science 268:714–716

Harvey RP, Degryse E, Stefani L, Schamber F, Cazenave JP, Courtney M, Tolstoshev P, Lecocq JP (1986) Cloning and expression of cDNA coding for anticoagulant hirudin from blood-sucking leech, Hirudo medicinalis. Proc Natl Acad Sci USA 83:1084–1088

Hein MB, Yeo T-C, Wang F, Sturtevant A (1996) Expression of cholera toxin subunits in plants. In: Collins GB, Shepherd RJ (eds) Engineering plants for commercial products and applications. New York Academy of Sciences, New York, pp 50–56

Herbers K, Wilke I, Sonnewald U (1995) A thermostable xylanase from Clostridium thermocellum expressed at high levels in the apoplast of transgenic tobacco has no detrimental effects and is easily purified. Biotechnology 13:63–66

Higo K, Saito Y, Higo H (1993) Expression of a chemically synthesized gene for human epidermal growth factor under the control of cauliflower mosaic virus 35S promoter in transgenic tobacco. Biosci Biotechnol Biochem 57:1477–1481

Hills MJ, Watson MD, Murphy DJ (1993) Targeting of oleosins to the oil bodies of oilseed rape. Planta 189:24

Horsch RB, Fry JE, Hoffmann NL, Eichholtz D, Rogers SG, Fraley RT (1985) A simple and general method for transferring genes into plants. Science 227:1229–1231

Jenkins N, Parekh RB, James DC (1996) Getting the glycosylation right: implication for the biotechnology industry. Nat Biotechnol 14:975–981

Kay R, Chan A, Dal M, McPherson J (1987) Duplication of CaMV 35S promoter sequences creates a strong enhancer for plant genes. Science 236:1299–1302

Kakkis ED, Matynia A, Jonas AJ, Neufeld EF (1994) Overexpression of the human lysosomal enzymes α-L-iduronidase in Chinese hamster ovary cells. Prot Express Pur 5:225–232

Kakkis ED, McEntee MF, Schmidtchen A, Neufeld EF, Ward DA, Gompf RE, Kania S, Bedolla C, Chien SL, Schull RM (1996) Long-term and high-dose trials of enzyme replacement therapy in the canine model of mucopolysaccharidosis I. Biochem Mol Med 58:156–167

Klocking H (1991) Toxicology of hirudin. Sem Thromb Hemostas 17:126–129

Kühnel B, Holbrook LA, Moloney MM and van Rooijen GJH (1996) Oil bodies of transgenic *Brassica napus* as a source of immobilized β-glucuronidase. J Am Oil Chem Soc 73:1533–1538

Kumagai MH, Turpen TH, Weinzattl N, Della-Cioppa G, Turpen AM, Donson J, Hilf ME, Grantham GL, Dawson WO, Chow TP, Piatak MJ, Grill LK (1993) Rapid high-level expression of biologically active α-trichosanthin in transfected plants by an RNA viral vector. Proc Natl Acad Sci USA 90: 427–430

Lehman ED, Joyce JG, Bailey FJ, Markus HZ, Schultz LD, Dunwillei CT, Jacobson MA, Miller WJ (1993) Expression, purification and characterization of multigram amounts of a recombinant hybrid HV1-HV2 hirudin variant expressed in *Saccharomyces cerevisae*. Protein Expr Pur 4:247–255

Lindsey K (1996) Plant transformation systems. In: Owen MRL, Pen J (eds) Transgenic plants: a production system for industrial and pharmaceutical proteins. Wiley, Chichester, pp 5–26

Loison G, Gindeli A, Bernard S, Nguyen-Juilleret M, Marquet M, Riehl-Bellon N, Carvallo D, Guerra-Santos L, Brown SW, Courtney M, Roitsch C, Lemoin Y (1988) Expression and secretion in *S. cerevisiae* of biologically active leech hirudin. Biotechnology 6:72–77

Lui DTY (1992) Glycoprotein pharmaceuticals – scientific and regulatory considerations and the United States Orphan Drug Act. Trends Biotechnol 10:114–120

Ma JK-C, Hiatt A, Vine ND, Wand P, Stabila P, van Dollerweerd C, Mostov K, Lehner T (1995) Generation and assembly of secretory antibodies in plants. Science 268:716–719

Ma S, Zhao D, Yin A, Mukherjee R, Singh B, Qin H, Stiller CR, Jevnikar AM (1997) Transgenic plants expressing autoantigens fed to mice induce oral immune tolerance. Nat Med 3:793–796

Markwardt F, Hauptmann J, Nowak G, Kleben C, Walsmann P (1982) Pharmacological studies on the antithrombotic action of hirudin in experimental animals. Thromb Haemostasis 47:226–229

Mason HS, Lam D, Arntzen CJ (1992) Expression of hepatitis B surface antigen in transgenic plants. Proc Natl Acad Sci USA 89:11745–11749

Matsumoto S, Ikura K, Ueda M, Sasaki R (1995) Characterization of a human glycoprotein (erythropoietin) produced in cultured tobacco cells. Plant Mol Biol 27:1163–1172

Miele L (1997) Plants as bioreactors for biopharmaceuticals: regulatory considerations. Trends Biotechnol 15:45–50

Moloney MM and van Rooijen GJH (1996) Recombinant proteins via oleosin partitioning. Inform 7:107–113

Moskowitz SM, Dlott B, Chuang PD, Neufeld EF (1992) Cloning and expression of a cDNA encoding the human lysosomal enzyme α-L-iduronidase. FASEB J 6:A77 (abstract)

NIH Technology Assessment Panel on Gaucher Disease (1996) Gaucher disease: current issues in diagnosis and treatment. JAMA 275:548–553

Neufeld EF (1991) Lysosomal storage diseases. Annu Rev Biochem 60:257–280

Neufeld EF, Muenzer J (1989) The mucopolysaccharidoses. In: Scriver CR, Beaudet AL, Sly WS, Valle D (eds) The metabolic basis of inherited disease 6th edn. McGraw-Hill, New York, pp 1565–1587

Neufeld EF, Muenzer J (1995) The mucopolysaccharidoses. In: Sciver CR, Beaudet AL, Sly WS, Valle D (eds) The metabolic and molecular bases of inherited disease 7th edn. McGraw-Hill, New York, pp 2465–2494

Owen MRL, Pen J (1996) Transgenic plants: a production system for industrial and pharmaceutical proteins. Wiley, Chichester

Parmenter DL, Boothe JG, Van Rooijen GJH, Yeung EC, Moloney MM (1995) Production of biologically active hirudin in plant seeds using oleosin partitioning. Plant Mol Biol 29:1167–1180

Pen J (1996) Comparison of host systems for the production of recombinant proteins. In: Owen MRL, Pen J (eds) A production system for industrial and pharmaceutical proteins. Wiley, Chichester, pp 149–168

Pen J, Verwoerd TC, van Paridon PA, Buedeker RF, van den Elzen PJM, Geerse K, van der Klis JD, Versteegh JAJ, van Ooyen AJJ, Hoekema A (1993) Phytase-containing transgenic seed as a novel feed additive for improved phosphorus utilization. Biotechnology 11:811–814

Roberts GW, James S (1996) Prion diseases: transmission from mad cows? Curr Biol 6:1247–1249

Rydel TJ, Ravichandran KG, Tulinsky A, Bode W, Huber R, Roitsch C, Fenton JW II (1990) The structure of a complex of recombinant hirudin and human α-thrombin. Science 249:277

Scott HS, Nelson PV, Hopwood JJ, Morris CP (1991) Human α-L-iduronidase: cDNA isolation and expression. Proc Natl Acad Sci USA 88:9695–9699

Sijmons PC, Dekker BMM, Schrammeijer B, Verwoerd TC, van den Elzen PJM, Hoekma A (1990) Production of correctly processed human serum albumin in transgenic plants. Biotechnology 8: 217–221

Stone SR, Maraganore JM (1993) Hirudin and hirudin-based peptides. Methods Enzymol 223:312–336

Taylor CB (1997) Comprehending cosuppression. Plant Cell 9:1245–1249

Turpen TH, Reinl SJ, Charoenvit Y, Hoffman SL, Fallarme V, Grill LK (1995) Malarial epitopes expressed on the surface of recombinant tobacco mosaic virus. Biotechnology 13:53–57

Tzen JTC, Huang AHC (1992) Surface structure and properties of plant seed oil bodies. J Cell Biol 117:327–335

Usha R, Rohl JB, Spall VE, Shanks M, Maule AJ, Johnson JE, Lomonossoff GP (1993) Expression of an animal virus antigenic site on the surface of a plant virus particle. Virology 197:366–374

van Rooijen GJH, Moloney MM (1995a) Structural requirements of oleosin domains for the subcellular targeting to the oil body. Plant Physiol 109:1353–1361

van Rooijen GJH, Moloney MM (1995b) Plant seed oil-bodies as carriers for foreign proteins. Biotechnology 13:72–77

Vandekerckhove J, Van Damme J, Van Lijsebettens M, Botterman J De Block M, Vandewiele M De Clercq A, Leemans J, Van Montagu M, Krebbers E (1989) Enkephalins produced in transgenic plants using modified 2\S seed storage proteins. Biotechnology 7:929–932

Varki A (1993) Biological roles of oligosaccharides: all of the theories are correct. Glycobiology 3:97–130

Vaughan P (1996) Creutzfeld-Jakob disease–latest unknown in struggle to restore faith in the blood supply. Can Med Assoc J 155:565–568

Verwoerd TC, Pen J (1996) Phytase produced in transgenic plants for use as a novel feed additive. In: Transgenic plants: a production system for industrial and pharmaceutical proteins. Wiley, Chichester, pp 213–225

von Schaewen A, Sturm A, O'Neill J, Chrispeels MJ (1993) Isolation of a mutant *Arabidopsis* plant that lacks *N*-acetylglucosaminyltransferase I and is unable to synthesize Golgi-modified complex *N*-linked glycans. Plant Physiol 102:1109–1118

Weissenborn DL, Grabau EA, Pedersen K, Oishi KK, Bennett S, Velander W, Cramer CL (1995) Production of human protein C in transgenic tobacco. International Conference on Engineering Plants for Commercial Products and Applications, Lexington KY, October, 1995 (abstract 50)

Zhao HG, Li HH, Bach G, Schmidtchen A, Neufeld EF (1996) The molecular basis of San Filippo syndrome type B. Proc Natl Acad Sci USA 12:6101–6105

Zhu Z, Hughes KW, Huang L, Sun B, Lui C, Li Y, Hou Y, Li X (1994) Expression of human α-interferon cDNA in transgenic rice plants. Plant Cell Tissue Org Cult 36:197–204

Feasibility of Antibody Production in Plants for Human Therapeutic Use

D.A. Russell

1 Introduction

In the 24 years since the initial publication describing monoclonal antibodies (KOHLER and MILSTEIN 1975) the field has advanced in the manipulation, production, and use of such biologics. While the molecular cloning of the antibody coding regions and advances in immunology have helped direct the engineering of antibodies, it has also paved the way for alternate expression systems. Given the efficiency of agricultural production systems, and successes in commercialization of transgenic crops, it is not surprising that plants have been investigated for such production. The use of plants to replace more traditional mammalian cell production is based on both systems having similar protein synthesis machinery, including many posttranslational processing steps. However, for commercial production of antibodies to the standards necessary for therapeutic use, any new system must be accurate, reproducible, and competitive. This chapter discusses the potential of therapeutic antibody production in transgenic plants to meet those

Integrated Protein Technologies/Agracetus Campus, Monsanto Co., 8520 University Green, Middleton, WI 53562, USA criteria. The concepts described have been applied to the first human clinical trial for an injectable therapeutic from transgenic plants.

2 Antibody Diversity and Biosynthesis

Advances in molecular immunology have helped define the requirements for antibody-type products, and develop the tools to meet those challenges. A brief overview of immunoglobulin structure and function will help to orient the reader new to this field. As shown in Table 1, human antibodies fall into five major classes, each with functionally and structurally distinct features. Most therapeutic work to date has focused on the immunoglobulin G (IgG) class. As shown in Fig. 1, they are composed of two heavy chains and two light chains, with the structure maintained by inter- and intramolecular cysteine bridges. When produced by the immune system, these IgG antibodies are used to sequester and tag foreign agents for destruction. The mechanisms of this destruction, antibody-dependent cellular cytotoxicity (ADCC) and complement-mediated cytotoxicity (CMC), are dependent on the structure of the constant (nonvariable) central portion of the heavy-chain

Table 1. Summary of human immunoglobulin classes (from HAYNES and FAUCI 1994; ROITT 1994)

Human immunoglobulin classes	Gene products per Ig	Ig mass, (kDa)	Sera $t_{1/2}$ (days)	Primary location	Sera mg/l	Characteristics
IgG	2	150	23	Sera	10,000	Internal protection from organisms, toxins
IgG1					6,500	Protein A binding, placenta transfer, CMC, ADCC
IgG2					2,300	Protein A binding, poor placenta transfer, cmc, lower ADCC
IgG3					800	No protein A binding, placenta transfer, CMC, ADCC
IgG4					400	Protein A binding, placenta transfer, poor CMC, poor ADCC
IgA	4	385	5.8	Secretions	2,000	Mucosal protection from antigens
IgM	3	950	5.1	Sera	1,000	Free form binds antigens; bound as activator of B cells
IgD	2	175	2.8	Lymphocytes	400	Bound form in B cell activation
IgE	2	190	2.5	Mast cells	<0.5	Activates inflammation response to antigen

CMC, Complement-mediated cytotoxicity; ADCC, antibody-dependent cellular cytotoxicity

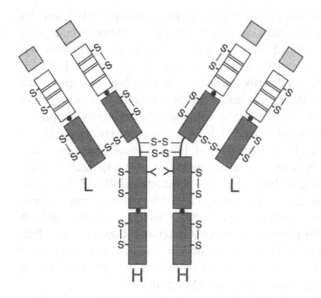

Fig. 1. Schematic structure of IgG molecule. Signal peptide cleaved from the mature N-terminus is shown (*stippled*). The variable domains of the light (kappa; *L*) and heavy (gamma; *H*) chains are shown with the three complementary determining regions, followed by the constant regions (*gray*). The conserved asparagine-linked glycosylation site in the CH$_2$ domain is shown (*Y*). (From MALONEY et al. 1995; ROITT 1994; HAYNES and FAUCI 1994)

(CH$_2$) region). Each subclass of IgG has a signature structure of the heavy-chain constant region, leading to differences in ADCC, CMC, and placenta transfer (HAYNES and FAUCI 1994). The CH$_2$ region of the heavy chain also is responsible for *Staphylococcus aureus* protein A binding. This binding has been exploited in the purification of IgG at the research and production scale (YOUNG et al. 1997; CARTWIGHT 1994). It is not dependent on glycosylation (NOSE and WIGZELL 1983).

Different mammals can produce homologous antibody classes and subclasses. Although they may be functionally similar, cross-species administration can invoke an immune response, due to amino acid sequence variation between the species-specific constant regions. Though mouse-generated monoclonal antibodies are presently clinically valuable as parenterals, the potential for a human anti-mouse antibody response (HAMA) must be monitored (MALONEY et al. 1995; ÖSTBERG and QUEEN 1995). To help minimize this potential, the constant region of the antibody can be replaced with the host-identical sequence, to generate a chimeric antibody (TRAIL et al. 1993).

Much of the constant region can also be eliminated by the genetic or enzymatic elimination of this region, to generate a Fab (antibody fragment). In a further derivation, the variable region of the heavy and light chain can be joined by a linker peptide, to generate a single-chain variable fragment (scFv). Both derivatives retain antigen binding, but lose effector functions. A recent derivation of scFv's with two different variable domains (diabodies) may circumvent this limit (HOLLINGER et al. 1997). In addition to eliminating unneeded protein portions, the scFv also has the increased potential for synthesis in *E. coli*, where production costs can be lower. An scFv is also amenable to gene fusion, allowing the coupling of antigen binding with a novel toxic activity for targeted therapeutics (BRINKMAN et al. 1991; FRANCISCO et al. 1997). Genetic fusions are also possible with IgG (HORNICK et al. 1997).

As indicated by its biology described in Table 1, IgA has been of interest as an oral therapeutic agent. It is the normal immunoglobulin used for such protection of body mucosa (KERR 1990); its presence in milk may protect infants via passive immunity (GOLDMAN and GOLDBLUM 1995). In vitro, it can show better stability to protease degradation than IgG (REINHOLDT et al. 1990). However, IgG can protect mice from oral infection by *Streptococcus mutans* (MA et al. 1989), nasal infection by Sendai virus (MAZANEC et al. 1992) or vaginal infection by HSV-2 (ZEITLIN et al. 1998). A mouse IgG1 monoclonal against *E. coli* K99 delivered orally can minimize disease progression in calves (SHERMAN et al. 1983). A comparison of recombinant IgA vs. IgG showed similar protection from viral infection (MAZANEC et al. 1992). The value of IgA engineering of IgG molecules for improved human oral and enteric pharmacology will best be tested by class-switching of a specific IgG to an IgA, followed by in vivo evaluation over time (MA et al. 1994).

The IgG structure shown in Fig. 1 has several other details important to understand for effective IgG production in plants. Since all IgGs are secreted proteins, they require a signal peptide for targeting to the endoplasmic reticulum, the first organelle of the secretory system. Significant antibody or scFv accumulation is not observed without a signal peptide (HIATT et al. 1989; FIEDLER and CONRAD 1995). As the activities of these targeting sequences are well conserved across species, effective signal peptides can be used from yeast (HEIN at al. 1991), plants (DE NEVE et al. 1993; FRANCISCO et al. 1997), and mammals (HIATT et al. 1989). The general rules of signal peptide design are somewhat understood (von HEIJNE 1986). Coincident with import, the signal peptide is removed. Certain amino acid changes should be avoided in the region of the signal peptide processing site, though, since some variants can lead to poor signal peptide removal, protein expression, or folding (GILLIKIN et al. 1997).

The region on the amino terminal end of both the heavy and light-chain is the variable domain, responsible for antigen binding. Although the amino acid sequence of this domain of a mouse monoclonal IgG would be defined by hybridoma selection for antigen binding, changes can be made to preserve the antigen binding domain, yet minimize HAMA. This humanization process can lead to a less severe immune response to the therapeutic antibody (MALONEY et al. 1995; ÖSTBERG and QUEEN 1995). A more extreme possibility is the generation of human monoclonals, by immunizing genetically modified mice that contain only genes for human antibody production (LONBERG et al. 1994), or molecular cloning of human antibody cDNAs (BENDER et al. 1994).

Coincident with translocation into the endoplasmic reticulum lumen, disulfide bonds are formed by protein disulfide isomerase, and domains are folded, with the assistance of molecular chaperones. These classes of proteins and activities are found in all eukaryotes. As with other organisms, these activities in plants may be necessary for correct folding of the secreted proteins, or elimination of the misfolded protein via degradation (BERGERON et al. 1994). Upon successful completion of the above processes, the minimal biosynthesis has occurred to produce a viable antibody molecule.

3 Plant vs. Mammal Asparagine-Linked Glycosylation

Another significant posttranslational modification of antibodies and other secretory proteins is the addition of a high mannose glycan at specific asparagine (N-) residues. This addition occurs as a preformed structure, transferred from dolichol phosphate to the asn-X-ser/thr motif present in the translocating protein chain. As the antibody proceeds through the endoplasmic reticulum, and the golgi stacks, sugars are added and subtracted in a sequential trimming process. The initial glycan transferred to the antibody is identical in mammalian cells and transgenic plants. However, as shown in Fig. 2, the resultant trimming can yield a very different structure. Roughly 10% of the IgG from normal human sera contains one terminal neuraminic acid per CH_2 region. This addition occurs by a specific glycosyl transferase for each branch. The β-1,6 fucose linkage to the asparagine-linked glucosamine is more prevalent (DWEK 1995). Neither of these types of linkages are

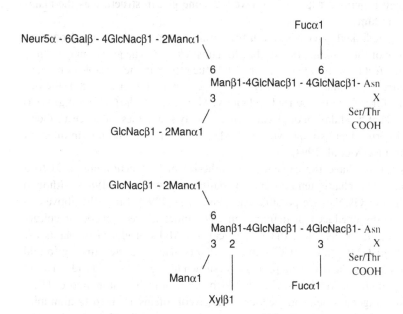

Fig. 2. Model asparagine-linked glycan structure from mammals (*above*) and plants (*below*). *Fuc*, Fucose; *Gal*, galactose; *GlcNac*, N-acetyl glucosamine; *Man*, mannose; *Neur*, neuraminic acid; *Xyl*, xylose. Mammalian structure can vary in presence of fucose, presence of bisecting (β1,4)GlcNac (not shown). Either branch may extend as far as NeurAc, in the order shown. Plant structure can vary in presence of fucose, presence of xylose. Either branch may terminate in GlcNac. Galactose is rarely seen in plants. (From DWEK 1995; FAYE et al. 1989; HAYASHI et al. 1990)

present in plants (FAYE et al. 1989). Plants also have very different rules for serine (O-) linked glycosylation.

The role of the terminal sugar residues on the antibodies has been tested by enzymatic removal of specific sugars, and by genetic or chemical perturbation of the process. The general conclusion is that terminal sugar pattern may be important for ADCC and CMC, but antibody stability may not be affected. Chemical inhibition of dolichol-glycan transfer to a mouse IgG2b monoclonal yielded antibody which maintained protein A binding, but lost ADCC and CMC potential, although in vivo clearance was similar (NOSE and WIGZELL 1983). When the conserved asparagine is changed to a nonaccepting glutamine, mouse human chimeric IgG1 showed a similar in vivo clearance rate ($t_{1/2}$ of 6.5 days), while the mutation of IgG3 led to a faster clearance rate ($t_{1/2}$ of 3.5 vs. 5.1 days). Both nonglycosylated molecules failed to bind complement (TAO and MORRISON 1989). In vitro protease sensitivity of IgA is greater after removal of the terminal neuraminic acid, but stability is enhanced over control when both neuraminic acid and N-acetyl galactosamine are removed from both the O- and N-linked carbohydrates (REINHOLDT et al. 1990). Mabs with terminal N-acetylglucosamine (GlcNac) tends to have decreased CMC activity, but not necessarily ADCC. A genetic mutant, creating glycans on an IgG1 Mab with mannose termini and lack of fucose, has reduced CMC activity and in vivo half-life (WRIGHT and MORRISON 1994). A similar mutant is available in the Arabidopsis plant (GOMEZ and CHRISPEELS 1994). A Mab produced in such a plant is expected to have the same glycan structure as the mammalian cell mutant.

Asparagine-linked glycosylation in the variable domain can occur, dependent on the nature of the shuffling of variable domains selected. The glycan may or may not have an effect on antibody behavior. While the glycan in the variable domain of an anti-dextran Mab increases affinity for antigen (WALLICK et al. 1988), the opposite is true for a set of humanized and mouse Mabs against the CD33 antigen (Co et al. 1993). That modulation of glycan structure may sometimes influence antibody binding is inferred from studies with an IgM Mab, from cells cultured in different sugars (TACHIBANA et al. 1994).

Plants do not have the capacity to synthesize or link neuraminic acid to a glycan chain. The chains tend to end in mannose (FAYE et al. 1989), although chains ending in GlcNac are possible (HAYASHI et al. 1990). From the above discussion of Mabs produced in mammalian cells under altered genetic or culture conditions, it can be assumed that the glycan on Mabs produced in plants are altered in their CMC and ADCC activities. Plants also have the capacity to add novel sugars during the trimming process. As shown in Fig. 2, fucose is added as an α-1,3 linkage, and a novel β-1,2-xylose is added to the branchpoint mannose. These residues are antigenic when endogenous plant glycoproteins are used to immunize rabbits (LAINE and FAYE 1988). Data to date on glycosylated Mabs produced in plants indicate the glycosylation is consistent with that of endogenous plant glycoproteins: little galactose detected, and significant mannose, GlcNac, xylose, and fucose are detected (D.A. Russell, unpublished data). This is consistent with the results of HEIN et al. (1991), who defined the glycosylation state of an antibody produced in tobacco cells by lectin affinity. However, limited data indicate the in vivo half-life is similar to that of a Mab produced in mammalian cells (ZEITLIN et al. 1998), and injections into mice of the plant-glycosylated antibody only succeeded in raising antibodies to the peptide portion of the antibody (D.A. Russell, unpublished data).

4 Market Requirements for Therapeutic Antibodies

While the diversity of the antibody variable domains creates a large mine of potential therapeutic agents, development of therapeutic Mabs to market has not been as easy as hoped for. Table 2 lists some of the current late stage monoclonal antibody products. The range of products reflects the utility of monoclonal antibodies and their derivatives. Although mouse monoclonal are current products, their repeated use can be limited by the potential HAMA response (MALONEY et al. 1995; ÖSTBERG and QUEEN 1995). Also represented are Fabs, chimeric, and humanized antibodies. Although antibody products from pooled human sera also are available, human monoclonals are not. There are also no present scFV products.

The range of products include a number of indications. Many are for cancer imaging: A radioactive tag is attached to the antibody, in order to follow the localization to any tumor cells, and aid in defining the course of therapy (ZUCKIER and DENARDO 1997). As the antibodies have affinity for surface antigens of cancer cells, some of the same antibodies are being evaluated as therapeutics, to target toxins to the cells. The toxins may be higher energy radioactive elements or other toxins (MALONEY et al. 1995; BRINKMAN et al. 1991). The example listed, Rituxan, is a chimera of a mouse variable domain, specific for a B-cell protein common in non-Hodgkin's lymphoma, and a human gamma 1 constant region. The antibody binds to cancerous as well as some noncancerous B-cells. The human constant region targets the cells for destruction via CMC and ADCC activities (ANDERSON et al. 1997). As noted above, these activities are influenced by the glycan structure.

Most examples shown are for use as parenterals. The one exception in this set is HNK20, a mouse IgA monoclonal. It was designed to prevent infection by Rous Sarcoma Virus in at-risk infants. To aid in patient compliance, it was delivered as nasal drops. Unfortunately, there was no clinically significant advantage seen in the first phase III clinical trials (DIBNER 1997). Greater success was seen with Medi-493, a humanized IgG used as a parenteral. These two outcomes also help illustrate the greater risk even late in pharmaceutical development, relative to that in agricultural biotechnology. The historic success rate for all biologics, from positive preclinical animal tests to product launch, is less than 40% (STRUCK 1996). While Table 2 reflects antibodies for human therapeutic use, there is interest in using the technology for mammalian therapies as well. One current product, Genecol® (Schering-Plough), is used to minimize newborn calf infection by E. coli K99. The mouse IgG antibody is purified from mouse ascites, and fed to the calf prophylactically. This

Table 2. Representative therapeutic mabs in late-stage development

Trademark/ Registration[a]	Product description	Source	Epitope	Use	1996 new US patients[b]	Dose	Status
Reopro	abciximab; 7E3 Fab	Centocor, Lilly	Glycoprotein IIb/IIIa	Anticlotting in angioplasty	398,000 (1993)	7.5mg/event	Marketed $540.02/10mg
Herceptin	trastuzumab; huMab	Genentech	Her-2	breast cancer	25% of 164,000	2-4mg/kg/week	Approved
Synagis	Medi-493; palivizumab; huMab	Medimmune	F-protein	RSV	90,000	15mg/kg per event	Approved
	HNK20; IgA	Oravax		RSV	90,000	2.4mg/day nasal	Phase III
Rituxan	Rituximab; IDEC-c2b8; huMab	Genentech, IDEC	CD-20	B cell non-Hodgkins Lymph.	50,000	375mg/m2 per week ×4	Marketed
Zenapax	Dacliximab; huMab	Protein Design Labs, Roche	IL-2 receptor, anti-TAC	Kidney transplant	12,003	1mg/kg/2 weeks	Marketed
Orthoclone OKT-3	Muromomab; CD3 Mab	Ortho Biotech	CD-3	Kidney, heart, liver transplant	12,003, 2342, 4062	70mg/ treatment	Marketed $672/5mg
CEA-scan arcitumomab	IMMU-4 technetium 99-Fab'	Immuno-medics, Mallinkrodt	Human 200 kDa CEA	Imaging colon, lung cancer	133,500, 177,000	1mg/event	Marketed
Prostascint	Capromab pendetide, cyt-356-I-111 Mab	Cytogen; CR Bard	100 kDa glycoprotein	Imaging prostate cancer	317,000	0.5mg/ event	Marketed
Oncoscint	Satumomab pendetide. I-111 Mab	Cytogen	TAG-72	Imaging ovary, colorectal	26,700. 133,500	1mg/event	Marketed $488/2mg
Verluma, Oncotrac	Mab-TC99; nofetumomab; Fab	NeoRX; DuPont-Merck	40-kDa glycoprotein	Imaging small cell lung can.	44,000	5-10mg/event	Marketed
OncoLym	Lym1-I131; Mab	Techiclone; Green Cross	HLA-DR variant Lym-1	Non-Hodgkin's lymphoma	50,000	≥5mg	Phase III

[a] MALONEY et al. 1995; ZUCKIER and DENARDO 1997; ANDERSON et al. 1996; CARDINALE 1997; PDR 1997; SHIELD et al. 1996; www.recap.com/mainweb.nsf/
[b] New United States patients from web sites www.cdc-gov/nchswww/releases; www.dcpc.nci.nih.gov/PLCO

passive immunization has also been investigated to protect other animals, such as chickens, from bacterial infection (WALLACH et al. 1990).

The volumes of product needed are dependent on real patient populations, and dosing volumes and schedules. Table 2 shows the new patient populations in the United States for the given indications. This estimate of patient need may be adjusted by a number of factors, such as length of therapy, relapses, and competing therapies. Some therapies are limited in repeat use: Abciximab, a mouse Fab fragment, may not be tolerated over long periods due to a HAMA response (PDR 1997). Antibody volume requirement also reflects therapeutic dose over the treatment period, and can vary greatly for the examples shown. In general, imaging agents use smaller amounts than therapeutic antibodies. CEA scan, if used for imaging once in all lung cancer patients, would require production of less than 200g antibody conjugate. Therapeutic antibodies used for the same indication have been at doses of up to 2g/treatment (MALONEY et al. 1995). Success would require production of as much as 177kg of antibody.

To date, all antibodies listed in Table 2 are produced from mammalian cell lines. Production levels of over 0.5g/l have been reported (BEBBINGTON et al. 1992; YOUNG et al. 1997), which is still a potential limit for high-volume antibody uses. Cost of production is estimated as US $300–1000/g, assuming yearly antibody needs of 100kg. Another system being tested, secretion into milk from transgenic animals, can produce 8g/l, with an estimated production cost of US $105/g (YOUNG et al. 1997). Modeling programs are available to estimate the costs (PETRIDES et al. 1995), both fixed (investments in fermentation facilities) and incremental (waste stream, analytics, personnel). This sort of analysis concludes that higher per unit costs are expected at lower volume needs, estimated at US $300/g for milk production at the 10kg scale (YOUNG et al. 1997). Although fast growing *E. coli* cannot make full antibodies, they have been used to make antibody derivatives, such as scFv's, at levels of 200mg/l (PACK et al. 1993). Any plant production system would have to compete with these systems. Cost of production is dependent on a number of factors, only some being directly related to the production system. Although early estimates of US $100/g for plant-produced antibodies were reported (HIATT 1990), the models may not have included all aspects of the process, validation, analytics, and infrastructure costs to a pharmaceutical substance, ready for formulation.

As late-stage development pharmaceuticals, all antibodies listed in Table 2 are at the status of phase III clinical trials or later. The stages of pharmaceutical development, in Table 3, define the path from initial gene discovery to marketing. The elapsed time on this path is dependent on many factors, and development can be stalled at any point (DIBNER 1997; STEIN 1997). As examples, Zenapax entered phase I clinical trials in 1992, phase II and III in 1994, and submission of a Biological Licence Application (BLA) in 1997. Phase II to BLA for Rituxan was 3.5 years. Any therapeutic antibody production system should not only meet volume and cost requirements, but these timelines as well. For comparison, it was 10 years from the first publication of a gene encoding tolerance to Roundup (KISHORE et al. 1986) to the sale of such an herbicide tolerant transgenic crop. In our work with transgenic corn, we have advanced from gene for transformation to

Table 3. Stages of drug development

Stage	Activities[a]	Time (years)[b]
Preclinical	In vitro activity, specificity; animal models; preparation for investigative new drug application	2.3
Phase I	Preliminary human safety and efficacy; wide dose, patients or healthy volunteers	1.8
Phase II	Targeted patient groups, efficacy and interactions	2.2
Phase III	Large-scale, multi-site testing	2
Biological License Application	Product description before sale: indications, Reactions, dose, manufacturing methods	2
Phase IV	Post-marketing safety and efficacy	

[a] FROELICH 1994.
[b] STRUCK 1996 (for all biologicals).

initial seed expression characterization within 8 months, and under 19 additional months for sustainable production from a defined, master seed bank (MSB) harvest. Part of the scaling is dependent on seed amplification rate, expression level, and total harvest mass. An inbred corn plant yields 100–500 seeds per 4–6 month generation, of 0.2g each. The variability is due to genetic line differences, as well as greenhouse vs. field growing conditions. Tobacco yields many logs more seed per plant, but at 11,000seeds/g (HANSON 1990). Corn is thus a better seed production system, but tobacco offers faster scaling times, and is useful for a leaf-derived product.

5 Plant Expression Systems

For plants to meet the critical endpoints outlined above in a competitive manner, the efficiencies of agricultural systems need to be tailored for the rigors of pharmaceutical production. Many of the points in the following sections have been implemented in our development of a corn seed system supporting human clinical trials. A useful starting point can be found in relevant FDA documents, available through the website www.fda.gov/cber/cberftp.html. Both "Points to Consider" and "Guidance" documents provide a framework in which to develop a plant pharmaceutical production system, with the goal of safe, effective, and consistent products. Much of the necessary infrastructure is foreign to the agricultural biotechnologist, but is a logical progression from current management practices with regulated transgenic plants found at www.aphis.usda.gov/biotech/ (USDA 1997a). This is similar to the progression from safe dairy practices to therapeutics production in milk of transgenic animals (FDA 1995). The detail required for approval of the Flavr Savr® tomato (REDENBAUGH et al. 1992), including molecular characterization, genetic stability, toxicology, and safety, is not too far removed from the requirements to characterize a pharmaceutical production system from a transgenic cell line or animal. Where the two paths diverge is in the processing of the raw material to a consistent drug substance (free of unwanted contaminants) compared to the entry of a deregulated transgenic plant into the food distribution system.

Many of the basic molecular tools and methods from agricultural biotechnology are directly applicable to meet the strict endpoints of pharmaceutical production in plants. As with mammalian and bacterial systems, higher expression levels can help reduce costs, due to lower throughput volumes in purification steps (PETRIDES et al. 1995; YOUNG et al. 1997). Although different mammalian host species, culture conditions, or transgene expression levels can influence posttranslational modification (WRIGHT and MORRISON 1994; YAN et al. 1990), these variables have not been reported in plant systems. For antibody production, much published data is in tobacco, using general gene expression elements, such as the CaMV 35S promoter, and nopaline synthase polyA addition site. These tools allow expression in most plant organs, although data are usually collected on leaves alone. As reviewed in CONRAD and FIEDLER (1994), antibody expression reported by different researchers varies from 0.004%–1.3% total soluble protein (tsp) in the leaves, with no obvious differences due to source of signal peptide. A range is also seen within the set of independent transformants created from one transgene (DENEVE et al. 1993). Such variation is also seen in mammalian systems (BEBBINGTON et al. 1992), and is presumably due to the variable nature of each integration event. Alternative plant expression systems may have some use in ensuring less variability. A recombinant TMV system is capable of high-level expression (DELLA-CIOPPA and GRILL 1996) but has not been show to create multi-subunit molecules, such as antibodies. Transformation of the plastid organelle, being a homologous recombination process, has less variability between independent events. Although very high expression levels have been achieved (MCBRIDE et al. 1994), the assembly of normally secreted and multi-subunit proteins such as antibodies has not been demonstrated. Success in site-specific integration to the nuclear genome should lead to less variability and higher expression success; it is being pursued in a number of systems, including plants (ALBERT et al. 1995).

Improvements in expression can be attempted by varying many of the genetic elements. In order to accelerate gene design, transient assays can be performed: DNA is delivered to cells, which are harvested for analysis one to three days later. By this technique, improved expression due to elements within the gene cassette have been demonstrated. These gene expression tests typically are carried out with model proteins, such as β-glucuronidase (CALLIS et al. 1987; NORRIS et al. 1993; FRISCH et al. 1995). Our own work has shown transient expression analysis can be applied to directly test antibody-encoding constructs. Shown in Fig. 3 is a nonreducing SDS-PAGE/western blot from both transiently and stably expressing corn tissue. In all crude plant samples, assembled antibody can be seen comigrating with purified protein. In addition, partially assembled molecules are seen in both crude plant extracts, indicating the transient system can model the processing success of the stable transformant. Through this test, we have been able to evaluate expression levels due to changes in introns, signal peptides, codons, and polyA addition sites, in order to prioritize constructs for stable transformation. In general, introns have

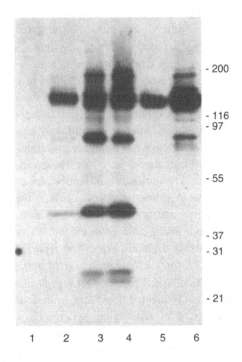

Fig. 3. Antibody accumulation from corn transient predicts stable expression. Antibody accumulation was analyzed by detection of kappa antigen via western blots of nonreducing SDS-PAGE. Fully assembled 150-kDa antibody is detected as well as some partially assembled products. *Lane 1*, immature corn embryo null control; *lane 2*, transient embryo expression; *lane 3*, mature transgenic embryo; *lane 4*, mature transgenic endosperm; *lane 5*, purified antibody from mammalian cells; *lane 6*, partially purified antibody from corn seed

the potential to increase expression in monocot plants (corn and rice; CALLIS et al. 1987) but less of an effect is seen in dicots (tobacco and arabidopsis; NORRIS et al. 1993). Any plant material amenable to transient analysis can be evaluated for antibody accumulation and assembly long before stable plants are made. This will help define the best species and target tissue for use in production. Transient assays can have limited value for comparison of tissue-specific promoters, possibly due to the structural difference between a free DNA transgene in a transient assay and a transgene integrated into the chromosome (FRISCH et al. 1995).

The use of tissue-specific promoters may allow higher expression, but must be active in the organ and species best suited for production. Tobacco for leaf harvest can scale-up quickly, due to high seed yields, and ease of clonal propagation. However, unless the protein of interest is stable in the harvested leaf during senescence and drydown (VERWOERD et al. 1995), an infrastructure to quickly process wet leaves would need to be developed. Seed expression allows the harvest and processing of a dried product, with a developed shipping, storage, and processing infrastructure. Potentially useful promoters include ubiquitin in corn seed (HOOD et al. 1997), and seed-storage promoters in corn endosperm (RUSSELL and FROMM 1997) and dicot seeds (FIEDLER and CONRAD 1995; FRISCH et al. 1995).

6 Plant Seeds as a Pharmaceutical Production System

With an interest in therapeutic proteins of greater than 10 kg volumes, we have been working with seeds as a production system. Seed production has the potential to capture the full value of a plant production system: flexibility in scaling, culturing, and processing. Storage and shipping of large quantities of seed allows flexibility not seen with many other systems. Simple storage at cool temperatures and controlled humidity allow viable seed recovery for at least 7 years (HANSON 1990). The full value of corn as a production system is realized when the diversity of germplasm is appreciated. This must be weighed with the efficiency of direct transformation of that material, or the speed to breed the antibody gene to the superior line. Lines can differ greatly in their transformation efficiency (ISHIDA et al. 1996), as can yield. Genetically purer inbred population yields are 0.5–2 MT/acre (M. Thompson, personal communication; where 1 MT = 39 bushels), and worsen with any inbreeding that might be done to quickly develop a line homozygous for the transgene (HALLAUER et al. 1988). Agronomic value is gained by crossing two genetically dissimilar corn lines, to generate high yielding hybrid plants in the next generation. Current hybrids can yield up to 3.3–5.3 MT/acre (GRAVES et al. 1996), although average 1996 yields in the United States were 3.1 MT/acre. For reference, average dry tobacco leaf harvest was 1 MT/acre (USDA 1997b).

As the raw harvested product should constitute a fraction of the cost of processing to a drug product (PETRIDES et al. 1995; YOUNG et al. 1997), the primary value of improved yield is in easier management of the crop for pharmaceutical production. Unfortunately, if a breeding strategy leads to only one parent carrying the antibody transgene, reduced gene expression may result in the initial, hemizygous seed product (RUSSELL and FROMM 1997). For antibody production in transgenic corn seed, Table 4 shows that expression is relatively stable across generations and field locations, but the gene dose may lower expression. Breeding the transgene into a second line to develop a hybrid system homozygous for the transgene, or into a better producing inbred that is not efficiently transformable, would require between three and ten generations, depending on the degree of genetic difference between the old and new line, and the degree of introgression desired (HALLAUER et al. 1988). This may be an unacceptable delay in product launch, but improvements might justify a replacement of the poorer producing line after launch.

Table 4. Expression is stable through generations, but is diluted with lower gene dose

Seed generation	Location	Gene dose	Total soluble protein (%)
R2 inbred	Greenhouse	3	3.0
R3 inbred	Field A	3	4.5
Hybrid	Field A	0–3	1.9
R3 inbred	Field B	3	3.3
Hybrid	Field B	0–3	1.6

The transgenic plant must be segregated so that it is genetically pure for one and only one drug product. Methods must be applied at every stage of production, from defining the initial material (master seed bank, MSB), to harvested production material (MIELE 1997). Fortunately, the isolation system required to maintain the genetic purity of field, sweet, and popcorn is readily adaptable to the needs of corn producing therapeutic proteins. For the most rigorous agricultural standards, that of foundation seed, isolation from other corn lines is necessary to ensure wind-born pollen does not contaminate the harvest. The typical restriction is to maintain mature plants of a single genotype isolated by 660 feet (distance separation). Additional control can be obtained by ensuring neighboring plants are without concurrently mature anthers and silks (temporal separation). The seed meets the genetic purity standard if no more than 1 of 1000 harvested seeds yields a plant of unexpected growth characteristics, or off-type (HANSON 1990). Soy and tobacco genetic isolation is somewhat easier, since the flowers tend to self-pollinate (WRUBEL et al. 1992).

Genetic purity due to mutation must also be assured. Although point mutations in a transgene leading to amino acid changes have been observed in mammalian cell culture (HARRIS et al. 1993), this has not yet been observed in transgenes of plants. As in any organism, though, spontaneous point mutations do occur in plants. Genes may also be changed by random insertions, such as retrotransposons in mammals (AMARIGLIO and RECHAVI 1993), or a variety of transposable elements in plants. When tested, most corn breeding lines have low transposon activity (CORMACK et al. 1988; DÖRING 1989), typically yielding a mutation rate similar to that seen in mammalian cells (STANKOWSKI et al. 1986). Plant transposon activity may increase during initial transformation steps (PESCHKE and PHILLIPS 1992). Part of MSB qualification would be to show the stability of the transgene by Southern analysis (REDENBAUGH et al. 1992) as well as sequencing (FDA 1996), irrespective of the cause of mutation.

Acceptance criteria of the harvested material for processing must also be determined and adhered to, in order to meet consistent therapeutic quality (FDA 1996). Application of agricultural food industry standards, a portion of which is shown in Table 5, provides an initial guide for these criteria. These reflect tolerances for edible products, not injectable therapeutics, so it is expected that purification will reduce these levels further. This concern is also reflected in the control of feed inputs for transgenic animals for therapeutics production (FDA 1995). The transgene product also must be stable during any shipping and storage period, in order to take full advantage of seed production. HOOD et al. (1997) have recently shown that avidin levels remain stable after 95 days of storage at 10°C. Production material must also be qualified in terms of disease vector material. This is most onerous for mammalian production hosts, where testing must occur for a variety of viruses and diseases, such as spongiform encephalopathy (FDA 1995, 1997). While plants do not propagate these agents, they do harbor a variety of bacteria and fungi. Part of process development should ensure that these agents are not in the final product. In addition, toxins from these organisms must also be reduced to acceptable levels. As shown in Table 6, we have been able to develop purification processes to minimize both classes of contaminants.

Table 5. Tolerances for some pesticide chemicals in popcorn and for fungal toxins (from Code of Federal Regulations, title 40, volume B, parts 150–189; www.access.gpo.gov/nara/cfr; PRICE et al. 1993)

Compound	Function	ppm
Diuron	Herbicide	1
Piperonyl butoxide	Insecticide	20
Pyrethrins	Insecticide	3
Methyl 3-[(dimethoxyphosphinyl) oxy]butenoate	Insecticide	0.25
Nicotine	Insecticide	2
Maneb	Fungicide	0.5
O,O-Diethyl S-[2-(ethylthio)ethyl] phosphorodithioate	Insecticide	0.3
Linuron	Herbicide	0.25
Dimethyl tetrachloroterephthalate	Herbicide	0.05
Butylate	Herbicide	0.1
Methomyl	Insecticide	0.1
3,4,5- and 2,3,5-Trimethylphenyl methylcarbamate	Insecticide	0.1
4-Aminopyridine	Bird repellent	0.1
3,5-Dimethyl-4-(methylthio)phenyl methylcarbamate	Pesticide	0.03
4-Amino-6-(1,1-dimethylethyl)-3-(methylthio)-1,2,4-triazin-5(4H)-one	Herbicide	0.05
Permethrin	Insecticide	0.05
Aflatoxin	Fungal product	0.02

Table 6. Endpoints and results of Mab purification from corn seed

Parameter	Initial	Final
Total gram Mab	38	23
Bioburden cfu g Mab	15,000	0
Endotoxin units/mg Mab	220,000	<0.1
HPLC purity	NA	>99%
HC sequence end	NA	MV
LC sequence end	NA	MD

7 Making a Drug Substance from Plants

With a qualified seed harvest, the challenge is to convert what appears to be an agricultural commodity into a drug substance, essentially equivalent to an antibody produced from any other system, and ready for formulation to drug product. The detail required at every step of this process is described as good manufacturing practices (GMP). The breadth of GMP compliance, from monitoring the air to waste stream, is described in KAPLAN (1995a,b). One aspect of validating a process for compliance is to add excessive amounts of a contaminant, in order to observe the degree and reproducibility of removing it during the process (CARTWRIGHT 1994; FDA 1997). This is done for a range of known contaminant types, which gives a comfort level for removing any new agents that may be discovered in the future.

The process developed to purify the therapeutic antibody must remove the contaminants listed in the previous section (Table 5), other plant proteins, and any interfering fragments of the antibody (FRANCISCO et al. 1997; MA et al. 1994).

Although all proteins and contaminants create unique challenges, some general principles apply. Methods must be shown to consistently meet purification end-points (FDA 1996). Using a diversity of purification steps may aid in removing different types of contaminants (CARTWRIGHT 1994). Once purified, the product needs to conform to standards, by both molecular and functional characterization (FDA 1997). Table 6 lists some of those endpoints and results for an antibody process from corn seed. By a number of parameters, the purified antibody is ef-fectively pure and homogeneous. This is somewhat surprising, when one considers that a typical fermentor run is for roughly 7 days under controlled environmental conditions (BEBBINGTON et al. 1992), while the corn seed has been accumulating the product for over 30 days under field conditions. Similar sequence homogeneity is seen with antibodies purified from tobacco leaves (HEIN et al. 1991) and avidin from corn seed (HOOD et al. 1997).

8 Summary

From the description above, the diversity of antibodies as a class of potential therapeutic agents is weighed against the constraints of developing any therapeutic molecule. Although much of this limit is specific to the antibody design, plant-based production systems have a potential to impact commercialization by making larger volume products manageable, with lower up-front capital requirements. Due to their novel glycosylation pattern (FAYE et al. 1989), plants at present may not create antibodies with all the functions of mammalian-glycosylated antibodies (WRIGHT and MORRISON 1994). This is not a limit for all current products. Success is dependent on fusing the efficient agriculture infrastructure with the narrow tol-erances required for a drug production system. Further validation of plants as a production system will come as more therapeutics from plants follow the corn-produced material through human clinical trials.

Acknowledgements. Special thanks to V. Paradkar for sharing of purification data, Cheryl Scadlock for assistance in computer searches, Agracetus members for developing the transgenic plants referred to in this review.

References

Albert H, Dale EC, Lee E Ow D (1995) Site-specific integration of DNA into wild-type and mutant *lox* sites placed in the plant genome. Plant J 7:649–659
Amariglio N, Rechavi G (1993) Insertional mutagenesis by transposable elements in the mammalian genome. Environ Mol Mutagen 21:212–218
Anderson DR, Grillo-López A, Varns C, Chambers KS, Hanna N (1997) Targeted anti-cancer therapy using rituximab, a chimaeric anti-CD20 antibody (IDEC-C2B8) in the treatment of non-Hodgkin's B-cell lymphoma. Biochem Soc Trans 25:705–708

Bebbington CR, Renner G, Thomson S, King D, Abrams D, Yarranton GT (1992) High-level expression of a recombinant antibody from myeloma cells using a glutamine synthetase gene as an amplifiable selectable marker. Biotechnology 10:169–175

Bender E, Pilkington GR, Burton DR (1994) Human monoclonal Fab fragments from a combinatorial library prepared from an individual with a low serum titer to a virus. Hum Antibodies Hybridomas 5(1–2):3–8

Bergeron JJM, Brenner MB, Thomas DY, Williams DB (1994) Calnexin: a membrane-bound chaperone of the endoplasmic reticulum. Trends Biochem Sci 19:124–128

Brinkman U, Pai LH, FitzGerald DJ, Willingham M, Pastan I (1991) B3(Fv)-PE38KDEL, a single-chain immutoxin that causes complete regression of a human carcinoma in mice. Proc Natl Acad Sci USA 88:8616–8620

Callis J, Fromm M, Walbot V (1987) Introns increase gene expression in cultured maize cells. Genes Dev 1:1183–1200

Cardinale V (1997) Drug topics redbook. Medical Economics, Montvale

Cartwright T (1994) Animal cells as bioreactors. Cambridge University Press, New York

Co MS, Scheinberg DA, Avdalovic NM, McGraw K, Vasquez M, Caron PC, Queen C (1993) Genetically engineered deglycosylation of the variable domain increases the affinity of an anti-CD33 monoclonal antibody. Mol Immunol 30:1361–1367

Conrad U, Fiedler U (1994) Expression of engineered antibodies in plant cells. Plant Mol Biol 26:1023–1030

Cormack JB, Cox DF, Peterson PA (1988) Presence of the transposable element Uq in maize breeding material. Crop Sci 28:941–944

Della-Cioppa G, Grill LK (1996) Production of novel compounds in higher plants by transfection with RNA viral vectors. In: Collins GB, Shepherd RJ (eds) Engineering plants for commercial products and applications. New York Academy of Sciences, New York

DeNeve M, DeLosse M, Jacobs A, VanHoudt H, Kaluza B, Weidle U, VanMatagu M, Depicker A (1993) Assembly of an antibody and its derived antibody fragment in Nicotiana and Arabdopsis. Transgenic Res 2:227–237

Dibner MD (1997) Biotechnology and pharmaceuticals: 10\years later. Biopharm 24–30

Döring HP (1989) Tagging genes with maize transposable elements: an overview. Maydica 34:73–88

Dwek RA (1995) Glycobiology: towards understanding the function of sugars. Biochem Soc Trans 23:1–25

Faye L, Johnson KD, Sturm A, Chrispeels MJ (1989) Structure, biosynthesis, and function of asparagine-linked glycans on plant glycoproteins. Physiol Plant 75:309–314

FDA (1995) Points to consider in the manufacture and testing of therapeutic products for human use derived from transgenic animals. United States Food and Drug Administration

FDA (1996) Guidance for industry in the submission of chemistry, manufacturing, and controls information for a therapeutic recombinant DNA-derived product or a monoclonal antibody product for in vivo use. United States Food and Drug Administration

FDA (1997) Points to consider in the manufacture and testing of monoclonal antibody products for human use. United States Food and Drug Administration

Fiedler U, Conrad U (1995) High-level production and long-term storage of engineered antibodies in transgenic tobacco seeds. Biotechnology 13:1090–1093

Francisco JA, Gawlak SL, Miller M, Bathe J, Russell D, Chace D, Mixan B, Zhao L, Fell HP, Siegall CB (1997) Expression and characterization of bryodin 1 and a bryodin 1-based single-chain immunotoxin from tobacco cell culture. Bioconjugate Chem 8:708–713

Frisch DA, van der Geest AHM, Dias K, Hall TC (1995) Chromosomal integration is required for spatial regulation of expression from the β-phaseolin promoter. Plant J 7:503–512

Froehlich J (1994) Phase I studies with biologicals. Eur. J Clin Res 6:124–129

Gillikin JW, Zhang F, Coleman CE, Bass HW, Larkins BA, Boston RS (1997) A defective signal peptide tethers the floury-2 zein to the endoplasmic reticulum membrane. Plant Physiol 114:345–352

Goldman AS, Goldblum RM (1995) Defense agents in milk. In: Robert G. Jensen (eds) Handbook of milk composition. Academic, San Diego

Gomez L, Chrispeels MJ (1994) Completion of an *Arabidopsis thalina* mutant that lacks complex asparagine-linked glycans with the human cDNA encoding N-acetylglucosaminyltransferase I. Proc Natl Acad Sci USA 91:1829–1833

Graves CR, Duck BN, Ellis FL, West DR, Kincer D, Thompson R, Percell G, Click CL, McClure J, Smith M, Williams JS (1996) Performance of corn hybrids from 1994 through 1996. University of Tennessee, Knoxville

Hallauer AR, Russell WA, Lamkey KR (1988) Corn Breeding. In: Sprague GF, Dudley JW (eds) Corn and corn improvement. American Society of Agronomy, Madison

Hanson AA (1990) Practical Handbook of agricultural science. CRC, Boca Raton

Harris RJ, Murnane AA, Utter SL, Wagner KL, Cox ET, Polastri GD, Helder JC, Sliwkowski MB (1993) Assessing genetic heterogeneity in production cell lines: detection by peptide mapping of a low level tyr to gln sequence variant in a recombinant antibody. Biotechnology 11:1293–1297

Hayashi M, Tsuru A, Mitsui T, Takahashi N, Hanzawa H, Arata Y, Akazawa T (1990) Structure and biosynthesis of the xylose-containing carbohydrate moiety of rice α-amylase. Eur J Biochem 191: 287–295

Haynes BF, Fauci AS (1994) Disorders of the immune system. In: Isselbacher KJ, Braunwald E, Wilson JD, Martin JB, Fauci AS, Kasper DL (eds) Harrison's principles of internal medicine. McGraw-Hill. New York

Hein MB, Tang Y, McLeod DA, Janda KD, Hiatt A (1991) Evaluation of immunoglobulins from plant cells. Biotechnol Prog 7:455–461

Hiatt A (1990) Antibodies produced in plants. Nature 344:469–470

Hiatt A, Cafferkey R, Bowdish K (1989) Production of antibodies in transgenic plants. Nature 342:76–78

Holliger P, Wing M, Pound JD, Bohlen H, Winter G (1997) Retargeting serum immunoglobulin with bispecific diabodies. Nat Biotechnol 15:632–636

Hood EE, Witcher DR, Maddock S, Meyer T, Baszczynski C, Bailey M, Flynn P, Register J, Marshall L, Bond D, Kulisek E, Kusnadi A, Evangelista R, Nikolov Z, Wooge C, Mehigh RJ, Hernan R, Kappel WK, Ritland D Li CP, Howard JA (1997) Commercial production of avidin from transgenic maize: characterization of transformant, production, processing, extraction and purification. Mol Breeding 3:291–306

Hornick JL, Khawli LA, Hu P, Lynch M, Anderson PM, Epstein AL (1997) Chimeric CLL-1 antibody fusion proteins containing granulocyte-macrophage colony-stimulating factor or interleukin-2 with specificity for B-cell malignancies exhibit enhanced effector functions while retaining tumor targeting properties. Blood 89:4437–4447

Kaplan R (1995a) Current GMP considerations for biotechnology facilities. I. Applicability of GMPs, technology transfer, and quality control. Biopharm 8:26–30

Kaplan R (1995b) Current GMP Considerations for biotechnology facilities. II. Process validation, contamination control, and compliance auditing. Biopharm 8:26–30

Kerr MA (1990) The structure and function of human IgA. Biochem J 271:285–296

Kishore GM, Brundage L, Kolk K, Padgette SR, Rochester D, Huynh K, Della-Cioppa G (1986) Isolation, purification and characterization of a glyphosate tolerant mutant E. coli EPSP synthase. Fed Proc 45:1506

Kohler G, Milstein C (1975) Continuous culture of fused cells secreting antibody of defined specificity. Nature 236:495–497

Laine AC, Faye L (1988) Significant immunological cross-reactivity of plant glycoproteins. Electrophoresis 9:841–844

Lonberg N, Taylor LD, Harding FA, Trounstine M, Higgins KM, Schramm SR, Kuo CC, Mashayekh R, Wymore K, McCabe JG et al (1994) Antigen-specific human antibodies from mice comprising four distinct genetic modifications. Nature 368:856–859

Ma JK, Hunjan M, Smith R, Lehner T (1989) Specificity of monoclonal antibodies in local passive immunization against Streptococcus mutans. Clin Exp Immunol 77:331–337

Ma JK-C, Lehner T, Stabila P, Fux CI, Hiatt A (1994) Assembly of monoclonal antibodies with IgG1 and IgA heavy chain domains in transgenic tobacco plants. Eur J Immunol 24:131–138

Maloney DG, Levy R, Campbell MJ (1995) Monoclonal antibody therapy. In: Mendelsohn J, Howley PM, Israel MA, Liotta LA (eds) The molecular basis of cancer. Saunders. Philadelphia

Mazanec MB, Lamm ME, Lyn D, Portner A, Nedrud JG (1992) Comparison of IgA versus IgG monoclonal antibodies for passive immunization of the murine respiratory tract. Virus Res 23:1–12

McBride KE, Schaaf DJ, Daley M, Stalker DM (1994) Controlled expression of plastid transgenes in plants based on a nuclear DNA-encoded and plastid-targeted T7 RNA polymerase. Proc Natl Acad Sci USA 91:7301–7305

Miele L (1997) Plants as bioreactors for biopharmaceuticals: regulatory considerations. Tibtech 15:45–50

Norris SR, Meyer SE, Callis J (1992) The intron of Arabidopsis thaliana polyubiquitin genes is conserved in location and is a quantitative determinant of chimeric gene expression. Plant Mol Biol 21:895–906

Nose M, Wigzell H (1983) Biological significance of carbohydrate chains on monoclonal antibodies. Proc Natl Acad Sci USA 80:6632–6636

Östberg L, Queen C (1995) Human and humanized monoclonal antibodies: preclinical studies and clinical experience. Biochem Soc Trans 23:1038 1043

Pack P, Kujau M, Schroeckh V, Knüpfer U, Riesenberg D, Plückthun A (1993) Improved bivalent miniantibodies, with identical avidity as whole antibodies, produced by high cell density fermentation of Escherichia coli. Biotechnology 11:1271 1277

Peschke VM, Phillips RL (1992) Genetic implications of somaclonal variation in plants. Adv Genet 30:41–75

Petrides D, Sapidou E, Calandranis J (1995) Computer-aided process analysis and economic evaluation for biosynthetic human insulin production – A case study. Biotechnol Bioengin 48:529 541

Price WD, Lovell RA, McChesney DG (1993) Naturally occurring toxins in feedstuff: Center for Veterinary Medicine Perspective. J Anim Sci 71:2556–2562

PDR (1997) Physician's desk reference. Medical Economics, Montvale

Redenbaugh K, Hiatt W, Martineau B, Kramer M, Sheehy R, Sanders R, Houck C, Emlay D (1992) Safety assessment of genetically engineered fruits and vegetables: a case study of the Flavr Savr tomato. CRC, Boca Raton

Reinholdt J, Tomana M, Mortensen SB, Kilian M (1990) Molecular aspects of immunoglobulin A1 degradation by oral streptococci. Infect Immun 58(5):1186–1194

Roitt IM (1994) Essential Immunology. Blackwell, Oxford

Russell DA, Fromm ME (1997) Tissue-specific expression in transgenic maize of four endosperm promoters from maize and rice. Transgenic Res 6:157–168

Sherman DM, Acres SD, Sadowski PL, Springer JA, Bray B, Raybouls TJ, Muscoplat CC (1983) Protection of calves against fatal enteric colibacillosis by orally administered Escherichia coli K99-specific monoclonal. Infect Immun 42:653 658

Shield CF III, Jacobs RJ, Wyant S, Das A (1996) A cost-effective analysis of OKT3 induction therapy in cadaveric kidney transplantation. Am J Kidney Dis 27:855–864

Stankowski LF Jr, Tindall KR, Hsie AW (1986) Quantitative and molecular analyses of ethyl methanesulfonate- and ICR 191-induced mutation in AS52 cells. Mutat Res 160:133 147

Stein KE (1997) Overcoming obstacles to monoclonal antibody product development and approval. Tibtech 15:88–90

Struck MM (1996) Vaccine R&D success rates and development times. Nat Biotechnol 14:591593

Tachibana H, Taniguchi K, Ushio Y, Teruya K, Osada K, Murakami H (1994) Changes of monosaccharide availability of human hybridoma lead to alteration of biological properties of human monoclonal antibody. Cytotechnology 16:151–157

Tao M, Morrison SL (1989) Studies of aglycosylated chimeric mouse-human IgG: role of carbohydrate in the structure and effector functions mediated by the human IgG constant region. J Immunol 143:2595–2601

Trail PA, Willner D, Lasch SJ, Henderson AJ, Hofstead S, Casazza AM, Firestone RA, Hellström I, Hellström KE (1993) Cure of xenografted human carcinomas by BR96-duxorubicin immunoconjugates. Science 261:212 215

USDA (1997a) Introduction of organisms and products altered or produced through genetic engineering which are plant pests or which there is reason to believe. United States Government Printing Office, Washington

USDA (1997b) Agricultural Statistics. United States Government Printing Office, Washington

Verwoerd TC, van Paridon PA, van Ooyen AJJ, van Lent JWM, Hoekema A, Pen J (1995) Stable accumulation of Aspergillus niger phytase in transgenic tobacco leaves. Plant Physiol 109:119 1205

von Heijne G (1986) A new method for predicting signal sequence cleavage sites. Nucleic Acids Res 14:4683–4690

Wallach M, Pillemer G, Yarus S, Halabi A, Pugatsch T, Mencher D (1990) Passive immunization of chickens against Eimeria maxima infection with a monoclonal antibody developed against a gametocyte antigen. Infect Immun 58:557 562

Wallick SC, Kabat EA, Morrison SL (1988) Glycosylation of a V_H residue of a monoclonal antibody against $\alpha(1\rightarrow6)$ dextran increases its affinity for antigen. J Exp Med 168:1099 1109

Wright A, Morrison SL (1994) Effect of altered CH_2-associated carbohydrate structure on the functional properties and in vivo fate of chimeric mouse-human immunoglobulin G1. J Exp Med 180:1087 1096

Wrubel RP, Krimsky S, Wetzler RE (1992) Field testing transgenic plants: an analysis of the US Department of Agriculture's environmental assessments. Bioscience 42:280 289

Yan SCB, Razzano P, Chao YB, Walls JD, Berg DT, McClure DB, Grinnell BW (1990) Characterization and novel purification of recombinant human protein C from three mammalian cell lines. Biotechnology 8:655–661

Young MW, Okita WB, Brown EM, Curling JM (1997) Production of biopharmaceutical proteins in the milk of transgenic dairy animals. Biopharm 10:34–38

Zeitlin L, Olmsted SS, Moench TR, Co MS, Martinell BJ, Paradkar VM, Russell DR, Queen C, Cone RA, Whaley KJ (1998) A humanized monoclonal antibody produced in transgenic plants for immunoprotection of the vagina against genital herpes. Nature BioTechnology 16:1361–1364

Zuckier LS, DeNardo GL (1997) Trials and tribulations: oncological antibody imaging comes to the fore. Semin Nucl Med 27:10–29

Pokeweed Antiviral Protein and Its Applications

N.E. TUMER, K. HUDAK, R. DI, C. COETZER, P. WANG, and O. ZOUBENKO

1 Introduction

The genus *Phytolacca* produces a number of proteins that have antiviral properties. These antiviral proteins are ribosome-inactivating proteins (RIPs) which remove a single adenine from a highly conserved, surface-exposed, stem-loop structure in the large rRNA of eukaryotic and prokaryotic ribosomes. They are found in two general forms: dimeric toxins (type II) containing a cell binding protein linked to the RIP by a disulfide bond, and single chain RIPs (type I), such as those found in pokeweed, composed of a single chain. A number of single-chain RIPs have been isolated from leaves, seeds and roots of a wide variety of plants (for reviews see IRVIN and UCKUN 1992; IRVIN 1995).

Biotechnology Center for Agriculture and the Environment and Department of Plant Pathology, Rutgers University, New Brunswick, NJ 08901-8520, USA

The presence of pokeweed antiviral protein (PAP) in *Phytolacca americana* was detected by its ability to inhibit the transmission of various plant viruses. Subsequently, PAP was found also to be effective against animal viruses, including HIV. In recent years, there has been a growing interest in using single-chain RIPs, such as PAP, as the cytotoxic component of immunotoxins targeted to cancer cells. PAP is an ideal candidate for therapeutic applications because it is well characterized and can be purified in relatively large amounts to prepare the quantities of immunotoxins required for clinical trials. PAP has been used as the cytotoxic moiety of immunotoxins against acute lymphoblastic leukemia and PAP-containing immunotoxins have shown significant antileukemic activity in clinical trials (for reviews see IRVIN and UCKUN 1992; IRVIN 1995).

RIPs are viewed as defense-related proteins because some RIPs, such as PAP, can deadenylate ribosomes from all organisms, and their expression in transgenic plants leads to resistance to viral infection. Expression of several RIPs is induced by environmental stress and it has been suggested that these RIPs may regulate protein synthesis during stress. Recent results demonstrated that expression of some RIPs, such as PAP in transgenic plants, leads to resistance to fungal infection (ZOUBENKO et al. 1997). Many other exciting applications of RIPs are likely to be developed in the near future.

2 Isolation and General Characteristics

PAP is easily purified from spring leaves of *Phytolacca americana* by ion chromatography and elution along a salt gradient. Leaves are homogenized in a neutral pH buffer and centrifuged to sediment remaining cellular fragments. The supernatant is fractionated between 60%–95% saturation of ammonium sulfate and the precipitate dialyzed against a low ionic strength buffer. The solution is passed through a DEAE cellulose column and PAP flows through in the unbound fraction that is then applied to a cation exchange resin S Sepharose column. The adsorbed PAP is eluted in a linear NaCl gradient. This procedure has been used to process 5kg leaves resulting in gram quantities of PAP with approximately 25% recovery of activity (IRVIN 1975; MYERS et al. 1991).

PAP is a basic protein with an isoelectric point of 9.5, a value that closely agrees with the estimated pI of 9.2 based on amino acid sequences (LIN et al. 1991). The protein has 28 basic amino acid residues that account for approximately 10% of the total amino acids and its basic pH. The molecular mass of PAP is 29kDa deduced by SDS-PAGE. Four cysteine residues establish disulfide bonds in the native PAP protein and their presence is considered to contribute to its noted thermal stability (KUNG et al. 1990). Reverse phase HPLC revealed multiple peaks, and analysis of two prominent peaks indicated an identical molecular mass, PI and antiviral activity. Recent results suggest that PAP may be synthesized by a multigene family as Southern blot analysis detected eight restriction fragments that hybridized to a cDNA probe containing the carboxyl-terminal end of the PAP gene (LIN et al. 1991). Therefore, the peaks resolved by the reverse phase HPLC may represent isozymes. RIPs have been isolated from both roots of pokeweed and cultured pokeweed cells with the same amino-terminal sequence and PI as PAP and therefore are likely PAP (BOLOGNESI et al. 1990; BARBIERI et al. 1989).

Other forms of PAP have been found, namely PAPII and PAP-S, although these have not been as extensively characterized as PAP. PAPII is isolated from leaves harvested in late summer and is purified in the same manner as PAP, but during the elution by NaCl gradient, elutes behind PAP at a higher salt concentration. The higher binding affinity of PAPII for the cation exchange column suggests that PAPII is a more basic protein than PAP. PAPII is also slightly larger, with a molecular mass of 30 kDa as estimated by SDS-PAGE (IRVIN et al. 1980) and does not cross-react with antibodies for PAP. The concentration of PAPII expressed in pokeweed varies and it has been suggested that PAPII production may be induced by environmental stress such as water deficit. The relationship between PAP and PAPII is a reflection of seasonal variation in expression of genes as the plant develops (HOUSTON et al. 1983).

PAP-S is expressed in pokeweed seeds and is more readily purified than PAP, by direct ion exchange chromatography of aqueous extracts of ground seeds (BARBIERI et al. 1982). The relative yield of PAP-S is higher than that of PAP from leaves, typically in the range of 100–180mg per 100g seeds. PAP-S comigrates with PAPII on SDS-PAGE even though estimates of its molecular mass, calculated from its amino acid sequence, indicate a protein of 29kDa. This discrepancy was explained by a later observation that PAP-S contains three N-acetylglucosamine residues, which raise its molecular mass to 29.8kDa (ISLAM et al. 1991). The PAP amino acid sequence has no glycosylation sites. The amino acid compositions of PAP, PAPII and PAP-S are very similar, with PAPII and PAP-S containing a higher proportion of basic amino acids as shown by their higher isoelectric points. Though PAPII does not cross-react with PAP antibodies, PAP-S does show partial cross-reaction and its enzymatic activity is inhibited in the presence of a five-fold increase in amount of antibody compared with that required to inhibit the activity of PAP (BARBIERI et al. 1982).

Comparison of the complete amino acid sequences for PAP and PAP-S indicate an 80% identity compared to 55% determined initially from the amino-terminal sequence. It was also discovered that PAP has a 31% identity with both ricin A-chain and trichosanthin and a 38% identity with saporin 6. Both trichosanthin and saporin 6, as with PAP, belong to the family of RIPs known as type I RIPs, all of which are single chain N-glycosidases.

Recently, the sequence of the complete cDNA encoding PAPII was determined and found to be only 33% homologous to PAP and PAP-S (POYET et al. 1994). The sequence encodes a protein of 285 amino acids and its expression in E. coli significantly diminished cell growth following its induction. Recombinant PAPII inhibited protein synthesis in a rabbit reticulocyte translation system and appeared to be as functional as native PAP in reducing protein synthesis in both eukaryotes and prokaryotes. The same group elucidated the complete cDNA encoding PAP-S and repeated the expression in *E. coli* to discover that, as with native PAP-S, recombinant PAP-S was more active than PAP in inhibiting eukaryotic in vitro translation. The PAP-S protein in seeds is glycosylated and even though recombinant PAP-S is not, it is expressed in *E. coli* in an active form, indicating that post-translational modification in pokeweed does not seem to alter its activity (POYET and HOEVELER 1997).

The internal sequence SEAARF, found at amino acids 175–180 in PAP, contains the catalytic amino acids characteristic of the active site of all ribosome-specific N-glycosidases. Three of these, the glutamic acid (E), the first alanine (A) and the arginine (R) are conserved in all. PAP-S contains 13 conserved residues found in the active site of 11 other RIPs including ricin, and 12 of these residues are common to PAP (KUNG et al. 1990; FUNATSU et al. 1991). MONZINGO et al. (1993) elucidated the three-dimensional structure of PAP and found it to be similar to ricin A-chain in folding pattern and the positions of key active site residues. PAP also complexed with formycin monophosphate, a substrate analog, more tightly than ricin A-chain. The formycin ring stacks between the invariant tyrosine residues, as it does in the ricin A-chain. The similarity in sequence and structure between the active sites of PAP and ricin suggests that their mechanisms of depurination are identical.

The cDNA sequence of PAP shows both amino- and carboxyl-terminal extensions on the native protein. These sequences are removed from the mature PAP and most likely direct the mature form to the extracellular matrix, and possibly protect ribosomes from the active protein. The amino-terminal leader of PAP contains 22 amino acids, the majority of which are nonpolar, suggesting that the protein is targeted outside the cell or across cellular membranes (LIN et al. 1991). The carboxyl extension of PAP contains 29 amino acids, the majority of which are also nonpolar, with two negatively charged residues near the carboxyl-terminus (LIN et al. 1991). Unlike the N-terminal extension, which is similar to the leader peptides of several other RIPs, the carboxyl-terminal extension of PAP bears no obvious similarities to other RIPs. Both dianthin 30 and saporin 6, type I RIPs, have N-glycosylation sites at this terminus, characteristic of proteins targeted to the vacuole (BENATTI et al. 1991; LEGNAME et al. 1991). Using immunogold cytochemistry, READY et al. (1986) showed that mature PAP is moved across cellular membranes and is localized in the cell wall matrix.

3 Enzymatic Function

3.1 Inhibition of Protein Synthesis

PAP was detected in the pokeweed plant in 1925 because of its ability to inhibit the transmission of various plant viruses (DUGGAR and ARMSTRONG 1925). In 1973, PAP was found to be an extremely potent inhibitor of protein synthesis in cell free extracts derived from eukaryotic organisms (IRVIN 1995). PAP damages the ribosome at sites where both elongation factors interact with the ribosome during the elongation cycle. PAP inhibits EF1 activity in two ways: by decreasing the factor-dependent binding of aminoacyl-tRNA to the acceptor site of the ribosome, and decreasing the ribosome-dependent hydrolysis of GTP mediated by this factor. However, the inhibition of enzymatic binding by PAP on ribosomes can be overcome by increasing the EF1 concentration (IRVIN et al. 1980). Since EF1 is extremely abundant in euckaryotic cells, and is present in large molar excess compared to the ribosomes and EF2 content, it is unlikely to be a rate limiting factor in protein synthesis. The major effect of PAP on protein synthesis is the inhibition of the translocation reaction mediated by EF2. The formation of the EF2-GDP-ribosome complex is inhibited by PAP. Direct measurement of the translocation reaction under defined conditions demonstrated a complete inhibition of translocation at catalytic concentrations of EF2 in the presence of PAP (GESSNER and IRVIN 1980). Treatment of posttranslocation ribosomes with PAP did not inhibit peptidyl transferase activity. PAP did not affect the initiation step of protein synthesis (RODES and IRVIN 1981).

3.2 Deadenylation of rRNA

In 1987, Endo and Tsurugi reported that ricin A-chain removed a single adenine from position 4324 in the 28 S rRNA of rat liver ribosomes, defining RIPs as ribosome-specific N-glycosidases. PAP depurinates intact ribosomes in exactly the same manner (BARBIERI et al. 1982; KUNG et al. 1990). Depurination occurs at a highly conserved stem-loop structure found in the large RNA of all ribosomes. The depurinated adenine is in the highly conserved sequence context of GAGA, shown to be involved in ribosome-elongation factor interaction. After depurination, the sugar phosphate backbone of the large rRNA is susceptible to hydrolysis and an RNA fragment can be seen in vitro, after incubation with acidic aniline.

The catalytic specificity of RIPs for rRNA may be modulated in a specific manner by the protein components of ribosomes. For example, some RIPs require an ATP-dependent factor to inactivate *Artemia salina* ribosomes (CARNICELLI et al. 1992), suggesting that at least in some cases, the need for cofactors may explain the variable effects of RIPs on ribosomes from different organisms.

3.3 Factors Influencing the Enzymatic Activity

Several type I RIPs, including PAP, are capable of depurinating both eukaryotic and prokaryotic ribosomes (HARTLEY et al. 1991) whereas the type II RIPs, such as ricin, have not yet been shown to affect prokaryotic ribosomes. Although ricin is inactive towards intact prokaryotic ribosomes, depurination of *E. coli* 23S rRNA by ricin has been described (ENDO and TSURUGI 1987). To investigate the reason for the differing ribosome specificity, CHADDOCK et al. (1996) created ricin/PAP hybrid proteins based on information from X-ray structures of ricin and PAP. Results indicated that the carboxyl-terminus of PAP (AA 219–262) does not contribute to ribosome recognition, whereas polypeptide changes in the amino-terminal half of the protein, within the first 126 residues, did affect ribosome inactivation. However, the differences in tertiary structure between ricin and PAP in their amino-termini were not responsible for altered ribosome specificity. Rather, more subtle changes in electrical charge within the regions 48–55 and 95–101 were found to determine ribosome recognition. Comparison of the electrostatic surfaces of PAP and ricin indicated that the charge organization on the surface of these proteins is different, particularly in the region of the amino-termini AA 48–55 and AA 95–101 (NICHOLLS et al. 1991). Therefore, it is likely that recognition of substrate by RIPs involves residues distant from the active site, in regions of the protein that may interact with ribosomal proteins to regulate specificity. It is unlikely that the active site itself determines recognition, as the key residues of all RIPs studied are conserved (ROBERTUS 1991), as are the ribosomal RNA target sequences. Also E. coli rRNA is depurinated by ricin in the absence of ribosomal proteins (ENDO and TSURUGI 1987).

Even though residues of the active site are conserved, recent work also points to differences in the structural requirement of the active site between type I and type II RIPs and sheds light on the recognition elements of PAP. The site of action of all RIPs is the universally conserved α-sarcin loop of E. coli 23 S rRNA, which is involved in the binding of elongation factors during protein translation (ENDO et al. 1988). RIPs act by removing a single adenine (A2660) from a GAGA-containing tetraloop structure at the apex of the α-sarcin loop. Interestingly, a point mutation that abolishes the tetraloop by disrupting the Watson-Crick base-pair involved in closing it, resulted in a loss of depurination of naked rRNA by ricin A-chain but not by PAP (MARCHANT and HARTLEY 1995). A double mutant, which restored the tetraloop, but in the reverse orientation, resulted in rRNA sensitivity to both PAP and ricin, as in the wild-type rRNA. Thus, the tetraloop structure is required for the enzymatic activity of ricin but not PAP. The action of PAP on E. coli ribosomes containing these mutants was the same as that on corresponding naked rRNAs. Given that the minimum substrate requirement for ricin action is a GAGA tetraloop attached by a helical stem of three base-pairs (GLUCK et al. 1992), it would appear that the recognition sites for ricin are contained within this structure. Therefore, the recognition elements for PAP and ricin differ and may account, in part, for the fact that PAP, unlike ricin, can catalyze the depurination of prokaryotic ribosomes. It is postulated that this finding may have an important bearing on conformational changes to the α-sarcin loop that occur during each elongation cycle. Perhaps in E. coli ribosomes, only the open, nontetraloop, configuration is accessible to action by RIPs and thus PAP, but not ricin, can catalyze depurination.

It was noticed that the efficiency with which PAP depurinated ribosomes varied depending on the reaction environment, suggesting that a cofactor was required for maximal enzyme activity. An 80-fold divergence in the IC_{50} values was reported for PAP acting on the unfractionated rabbit reticulocyte lysate translating endogenous mRNA compared with the translation of poly (U) in a system comprised of isolated ribosomes and elongation factors (CARNICELLI et al. 1992). These observations led to the finding that both ATP and tRNA, found in the postribosomal supernatant, were the cofactors that significantly upregulated the enzyme activity of PAP (BRIGOTTI et al. 1995). The addition of either ATP or tRNA alone did not result in this increase. Interestingly, the upregulation could be further enhanced by inclusion of the entire postribosomal supernatant, in addition to ATP and tRNA alone, suggesting that perhaps bound proteins are required to maintain the tRNAs in a more active configuration (UHLENBECK 1995).

The sensitivity of autologous ribosomes of pokeweed to PAP was initially challenged by the observation that PAP does not protect pokeweed from systemic infection by pokeweed mosaic virus (SHEPHERD et al. 1969). This led to the suggestion that perhaps a PAP-inhibitory agent existed in the cytosol of pokeweed cells. However, ribosomes combined with supernatant from pokeweed cells and incubated with PAP were as susceptible to depurination as ribosomes in the absence of supernatant (BONNESS et al. 1994), suggesting that PAP activity in pokeweed is not under the regulation of a cytosolic inhibitor.

3.4 Enzymatic Activity of PAP Against RNA and DNA

In addition to their activity on rRNA, in vitro activities of RIPs against both RNA and DNA have been described in recent years. RIPs such as trichosanthin, gelonin, ricin, luffin, cinnamomin and camphorin were shown to cleave supercoiled DNA into relaxed and linear forms (LI et al. 1991; HUANG et al. 1992; LING et al. 1994). A type I RIP, saporin L-1 was shown to be capable of releasing multiple adenines from a variety of RNAs, DNA and poly (A) (BARBIERI et al. 1994). As early as in 1985, OBRIG et al. had shown that ricin has RNase activity. Similarly, momorcharin, a RIP from *Momordica charantia*, was shown to degrade RNA nonspecifically (MOCK et al. 1996). Marchant and Hartley reported that PAP is capable of depurinating artificial polynucleotide loops which were not substrates for ricin (MARCHANT and HARTLEY 1995). Subsequently, some isoforms of PAP were shown to depurinate both ribosomal and nonribosomal substrates (STIRPE et al. 1996). NICHOLAS et al. (1997) reported that a type I RIP, gelonin degrades single-stranded DNA. BARBIERI et al. (1997) examined 52 purified RIPs, including both type I and type II RIPs, for ability to release adenine on various substrates including RNAs from different sources, DNA and polyA. All RIPs depurinated DNA and some released adenine from adenine-containing polynucleotides tested. Thus, a new name, polynucleotide: adenosine glycosidase, was suggested for RIPs (BARBIERI et al. 1997).

4 Cytotoxicity

The cDNA encoding PAP has been cloned, and in contrast to the healthy appearance of pokeweed, the expression of PAP in transgenic *N. tabacum* plants leads to various physiological changes. Transgenic tobacco plants producing high levels (more than 10ng/mg protein) of PAP were sterile with a stunted, mottled phenotype, which was correlated with the level of PAP expressed (LODGE et al. 1993). However, plants that produced less PAP (1–5ng/mg protein) were fertile and normal in appearance (LODGE et al. 1993).

A selection scheme used to isolate nontoxic PAP mutants takes advantage of the inability of *Saccharomyces cerevisiae* to grow in the presence of PAP. To select these mutants, *URA3* based yeast expression vectors were made using cDNAs encoding wild-type PAP and a variant of PAP (PAP-v), which contains two amino acid changes near the amino-terminus of the mature protein (LODGE et al. 1993). The expression of PAP from these vectors was under the control of the galactose inducible promoter, *GAL1*. In the presence of glucose, *GAL1* expression is repressed. In the presence of galactose, the *GAL1* promoter is induced, and growth is arrested due to expression of active PAP.

To isolate PAP mutants that can grow in the presence of galactose, the yeast expression plasmids pNT123 and pNT124 were mutagenized using hydroxylamine and transformed into yeast. These cells were plated on media containing glucose and replica plated onto galactose plates. Colonies growing on the galactose medium were streaked onto uracil-deficient galactose plates. The colonies that grew on uracil-deficient galactose plates were analyzed for PAP expression by ELISA and western blot analysis (HUR et al. 1995).

Several nontoxic PAP mutants were selected for sequence analysis. Some mutants comigrated with the precursor form of PAP (NT123-1 and NT124-1); another comigrated with the mature form of PAP (NT123-2); and others produced smaller proteins (NT124-2 and NT124-3). From the sequence analysis data these three classes of nontoxic PAP mutants are explained by (a) mutations near the amino-terminus that affect processing of the PAP precursor (NT123-1 and NT124-1), (b) an active site mutation (NT123-2); and (c) nonsense mutations leading to premature termination (NT124-2 and NT124-3).

For sequence analysis, plasmids expressing nontoxic PAP mutants were rescued from yeast and transformed into *E. coli*. Sequence analysis of NT123-1 indicated that there was a point mutation changing the glycine at position 75 to aspartic acid (G75 D). In NT124-1, a point mutation at position 97 changed a glutamic acid to a lysine (E97 K). The two proteins expressed from these mutant alleles comigrated with the precursor form of PAP, indicating that processing of PAP is inhibited by these mutations. The only change in NT123-2 was a glutamic acid to valine at position 176 (E176 V) at the active site. NT123-2 produced a protein the same size as wild-type PAP. NT124-2 and NT124-3 had identical point mutations near the carboxyl-terminus which altered a tryptophan at position 237 to a stop codon (W237). This nonsense mutation deleted 25 amino acid residues from the carboxyl-terminus of the mature PAP (HUR et al. 1995).

The X-ray crystal structure of PAP has been shown to be homologous to that of ricin-A chain (MONZINGO et al. 1993). There is a prominent fold in the three dimensional structure of PAP and ricin, corresponding to Glu 177, Arg 180, Trp 211 and Asn 209 of ricin. These residues have been shown to be conserved among several plant RIPs (LORD et al. 1994; MONZINGO and ROBERTUS 1992). Site directed mutagenesis by READY et al. (1991) showed that the negatively charged residue, Glu 177 is critical for ricin's N-glycosidase activity and that it plays a role in transition-state stabilization. One of the PAP mutants, NT123-2, had a mutation at Glu 176 (E176 V). The X-ray crystallography data shows that Glu-176 in PAP is at the equivalent position of Glu-177 in the active site of ricin (MONZINGO et al. 1993). The observation that this mutation, E176 V, makes PAP nontoxic to yeast supports the hypothesis that these glutamic acid residues are structurally and functionally equivalent.

In contrast to ricin mutations selected in yeast, which are predominantly at the active site (FRANKEL et al. 1989), changes to amino acid residues outside the active site of PAP reduced its inhibitory effect on yeast ribosomes. It is possible that these processing-defective PAP mutants might be less toxic due to a change in their compartmentalization within yeast. Alternatively, the mutation in NT123-1 (G75 D) occurs in a region of PAP that is predicted to form a beta sheet (MONZINGO et al. 1993). It is possible that this change could affect the conformation of this region of PAP such that orientation of the nearby Tyr 72 is altered, preventing Tyr 72 from binding to the target adenine base. A similar type of mutation R68G reported by CHADDOCK et al. (1994) reduced the activity of PAP to approximately 10% of the activity of wild-type PAP. R68G is located in the same β-sheet of PAP as our G75 V mutation. From X-ray analysis of PAP, R68 has the potential to hydrogen bond with Y72, a highly conserved active site residue. Molecular modeling experiments with RNA substrates and the active site structures of PAP and ricin A-chain indicate that Y72 (or Y80 in ricin) changes the orientation of the RNA backbone to allow efficient catalysis (CHADDOCK et al. 1994). Therefore, NT123-1 could lead to a misfolding of the PAP active site which would result in the loss of rRNA depurination activity.

Other mutants, NT124-2 and NT124-3 isolated from yeast contained nonsense mutations that lead to premature termination of the growing polypeptide chain. This results in the deletion of 25 amino acid residues from the carboxyl-terminus of the mature PAP. This carboxyl-terminal truncation removes a cysteine residue (Cys 259) which forms a disulfide bridge with an amino-terminal cysteine (Cys 34). Consequently, the disulfide bond between cysteine 34/259 is disrupted, leaving the second disulfide bond between cysteine 85/106 intact. Moreover, sequence analysis of PAP revealed that its carboxyl terminal sequence, EIKPDVALLNYVGGSC (residues 244–259), bears homology to the consensus sequence for the prokaryotic lipoprotein attachment site (HUR et al. 1995). In prokaryotes, the signal peptide of membrane lipoproteins is cleaved by signal peptidase II, which recognizes a conserved sequence, found in more than 50 prokaryotic proteins, and cuts after a cysteine residue to which a glyceride-fatty acid lipid is attached (HAYASHI and WU 1990). Prokaryotic membrane attachment sites are not correctly processed in eukaryotic cells, however, this homologous sequence may be important for PAP secretion. Therefore, the absence of this lipid binding sequence seems a more likely explanation for the lack of toxicity of NT124-2 and NT124-3 than the loss of a disulfide bond. This conclusion is supported by the finding that removal of a single disulfide bond in *Mirabilis* antiviral protein increases its inhibitory activity against rabbit reticulocyte ribosomes more than 20-fold (KATAOKA et al. 1991). The notion that the C-terminal region of PAP may be involved in its cytotoxity is further supported by the finding that addition of an endoplasmic reticulum retention sequence (KDEL) to the ricin A-chain significantly increases its cytotoxicity to mammalian cells (WALES et al. 1993). In this case, enhanced cytotoxicity was correlated with an increased level of ribosome inactivation, indicating that the added KDEL sequence facilitated ricin A-chain entry into the cytosol from the endoplasmic reticulum (WALES et al. 1993). The C-termini of reticuloplasmins, such as those of RIPs, share very little primary sequence homology. However, immunoprecipitation studies indicate that their C-termini fold into a common structure. It is possible that the same could be true for RIPs. These results indicate that the C-terminal regions of RIPs play an important role in their cytotoxicity and subcellular localization.

5 Antiviral Activity

5.1 Against Plant Viruses

The antiviral activity of PAP has two striking features, nonspecificity towards different viruses and strong antiviral activity at low concentrations. PAP inhibited local lesion formation by a number of different viruses, including both DNA and RNA viruses (CHEN et al. 1993). The concentration of PAP needed for inhibition of TMV infection in tobacco leaves was as low as 30nM (TAYLOR 1994). In tobacco protoplasts, complete inhibition of TMV was achieved at 300nM (GRASSO et al. 1979). PAP was the most effective antiviral protein among all RIPs tested (STEVENS 1981).

Transgenic plants, which expressed either PAP, or the double mutant derivative of PAP (PAP-v), showed resistance to infection by different viruses (LODGE et al. 1993). Resistance was effective against both mechanical and aphid transmission. Analysis of the vacuum infiltrate of leaves expressing PAP showed that it is enriched in the intercellular fluid in transgenic plants as it is in pokeweed (LODGE et al. 1993). Analysis of resistance in transgenic plants suggested that PAP confers viral resistance by inhibiting an early event in infection. Previous methods for creating virus resistant plants, such as pathogen derived resistance, have been specific for a particular virus or closely related viruses. In order to protect plants against more than one virus, multiple genes must be introduced and expressed in a transgenic line. Expression of PAP in transgenic plants indicated that resistance can be developed to a broad spectrum of plant viruses by expression of a single gene. In subsequent studies, virus resistance was observed with transgenic plants expressing other RIPs, such as trichosanthin (LAM et al. 1996) and dianthin (HONG et al. 1996).

As previously described, nontoxic PAP mutants were isolated by random mutagenesis and selection in yeast (HUR et al. 1995). One of these mutants, NT123-2, had a point mutation (E176 V) in the active-site that abolished enzymatic activity and another mutant, NT124-3, had a nonsense mutation that resulted in deletion of the C-terminal 25 amino acids (W237Stop). In vitro translation of rabbit reticulocyte lysate ribosomes was inhibited by the C-terminal deletion mutant, but not by the active-site mutant. When both mutants were expressed in transgenic tobacco, unlike PAP or the variant (PAP-v), neither mutant was toxic to transgenic plants (TUMER et al. 1997). In vivo depurination of rRNA was detected in transgenic tobacco expressing PAP-v, but not in-transgenic plants expressing either the active-site mutant or the C-terminal deletion mutant PAP. When extracts from transgenic plants containing the mutant PAPs were exogenously applied to tobacco leaves in the presence of potato virus X (PVX), the C-terminal deletion mutant had antiviral activity, while the active-site mutant had no antiviral activity. Furthermore, transgenic plants expressing low levels of the C-terminal deletion mutant showed resistance to PVX infection, while transgenic plants expressing very high levels of the active-site mutant PAP were not resistant to PVX. These results demonstrated that an intact active-site of PAP is necessary for antiviral activity, toxicity and in vivo depurination of tobacco ribosomes. These results also indicated that an intact active-site is not sufficient for all these activities. An intact C-terminus is also required for toxicity, and depurination of tobacco ribosomes in vivo, but not for antiviral activity, suggesting that antiviral activity of PAP can be dissociated from its toxicity (TUMER et al. 1997).

5.2 Against Animal Viruses

PAP has shown antiviral activity against a number of animal viruses including polio (USSERY and HARDESTY 1977), herpes simplex (TELTOW et al. 1983), influenza (TOMLINSON et al. 1974) and human immunodeficiency virus (ZARLING et al. 1990). The doses for inhibition of the different viruses varied depending on the types of cells used. PAP inhibited infection of both Vero and HeLa cells by herpes simplex virus (HSV) at micomolar concentrations (ARON and IRVIN 1980). PAP produced only a 30% inhibition of total protein synthesis in virus infected cells, but inhibited virus production greater than 90%. The studies with HSV also showed that HSV DNA synthesis was inhibited approximately 90% in PAP-treated cells with no effect on cellular DNA synthesis (ARON and IRVIN 1980). These results raised the possibility that the antiviral action of PAP may not be due to its inactivation of ribosomes.

PAP has been shown to have anti-HIV activity similar to trichosanthin. PAP inhibits production of p24 in both T cells and macrophages infected in vitro with an IC_{50} of 0.5nM (ZARLING et al. 1990). These studies also demonstrated that uninfected cells were not affected by PAP treatment up to 30 nM concentrations. PAP-S added at the time of infection was more effective than AZT in inhibiting the expression of reverse transcriptase in isolated mononuclear blood cells infected in vitro with HIV. The anti HIV activity of PAP was enhanced 1000-fold by conjugation to monoclonal antibodies reactive with different T-cell antigens, including CD4 (ZARLING et al. 1990). An anti-CD4-PAP immunoconjugate effectively reduced viral reverse transcriptase activity in zidovudine resistant T-cells. Studies using clinical isolates of zidovudine sensitive and resistant HIV-1 demonstrated that anti-CD4-PAP exhibited potent anti-HIV activity with IC_{50}s below 100 pM for all isolates (ERICE et al. 1993). This study also showed that anti-CD4-PAP immuno-toxin had no cytotoxic action against the lymphohematopoietic cell populations at concentrations effective against HIV infected T-cells. Anti-HIV activity of other RIPs has also been observed. Trichosanthin has a similar anti-HIV activity to PAP and now is in clinical trials. Other RIPs which possess anti-HIV activities are found in *Gelonium, Dianthus* and *Momordica* (HUANG et al. 1992; LEE-HUANG et al. 1994).

5.3 Against Yeast Viruses

Programmed ribosomal frameshifting is a molecular mechanism used by many RNA viruses to produce Gag-pol fusion proteins. The efficiency of these frameshift events determines the ratio of viral Gag to Gag-pol proteins available for viral particle morphogenesis, and changes in ribosomal frameshift efficiencies can se-verely inhibit virus propagation. Since ribosomal frameshifting occurs during the elongation phase of protein synthesis, it was hypothesized that agents that affect the different steps in this process may also impact programmed ribosomal frame-shifting. Two different viral systems (the L-A and M_1 "killer" system, and Ty*1*) of the yeast *S. cerevisiae* were utilized as models to study the antiviral activity of PAP (TUMER et al. 1998).

The ability of ribosomes to maintain the correct translational reading frame is fundamental to the integrity of translation and, ultimately, to cell growth and viability. Thus, the protein translational machinery has evolved to ensure that the intrinsic error rate of reading frame maintenance is extremely low, approximately 2×10^{-4} (THOMPSON and DIX 1982; MENNINGER 1975; DINMAN et al. 1991). How-ever, in a number of cases, elongating ribosomes are programmed to shift the translational reading frame. These instances of "programmed ribosomal frame-shifting" are most commonly observed in RNA viruses and play a critical biological role in viral particle morphogenesis (DINMAN 1995). The L-A Gag-pol fusion protein is produced as a consequence of a programmed −1 ribosomal frameshift event in which the elongating ribosome shifts the reading frame by 1 nucleotide to the 5' end of the mRNA. In Ty*1*, the Gag-pol fusion protein is produced when an elongating ribosome slips 1 nucleotide in the 3' direction, i.e., programmed +1 ribosomal frameshifting. There are also examples of programmed ribosomal frameshift events in a few bacterial cellular genes and at least one chromosomal eukaryotic gene (GESTELAND and ATKINS 1996). Programmed ribosomal frameshift events occur with a frequency of approximately 10^{-2}, two orders of magnitude more frequently than nonprogrammed frameshifts.

Altering the efficiency of −1 ribosomal frameshifting changes the ratio of Gag to Gag-pol, which in turn affects viral particle assembly and RNA packaging (DINMAN and WICKNER 1992). Increased efficiencies of +1 ribosomal frameshifting in polyamine-starved yeast cells correlated with decreased Ty*1* retrotransposition frequencies (BALASUNDARAM et al. 1994). PAP was shown to specifically inhibit Ty*l* directed + 1 ribosomal frameshifting in intact yeast cells and in an in vitro assay system (TUMER et al. 1998). Using an in vivo assay for Ty*l* retrotransposition, PAP was shown to specifically inhibit Ty*l* retrotransposition, suggesting that Ty*l* viral particle morphogenesis is inhibited in infected cells (TUMER et al. 1998). PAP did not affect programmed −1 ribosomal frameshifting efficiencies, nor did it noticeably impact the ability of cells to maintain the M1-dependent killer virus phenotype, suggesting that the −1 ribosomal frameshift does not occur after the peptidyl transferase reaction. These results provided the first evidence that PAP has viral RNA-specific effects in vivo which may be responsible for the mechanism of its antiviral activity (TUMER et al. 1998).

5.4 Antiviral Mechanism

Renewed attention has focused on the antiviral activity of PAP. Based on depurination activity and the extracellular location of PAP, a general antiviral mechanism was postulated (READY et al. 1986). According to this mechanism, PAP and other RIPs might enter the cell along with the virus at the site of infection and inhibit virus multiplication by inactivating host ribosomes. This hypothesis was supported by CHEN et al. (1993) who showed that there was a positive correlation between the depurination of tobacco ribosomes and the extent of inhibition of TMV infection. TAYLOR et al. (1994) also showed that barley RIP and ricin, which were not active on tobacco ribosomes, were not antiviral. However, as described above, antiviral activity and the extent of inhibition of host protein synthesis were not correlated well in the studies of PAP against animal viruses, suggesting that ribosome-inactivation might not be the only antiviral mechanism.

When cucumber mosaic virus was incubated with PAP, the virus regained full infectivity on *Chenopodium amaranticolor* when it was separated from PAP by centrifugation (TOMLINSON et al. 1974), suggesting that PAP does not irreversibly inactivate the viral particles. Further observations indicated that interaction between PAP and TMV was weak and could be disrupted at high salt concentrations (300 mM KCl), while the infectivity of TMV after it was separated from PAP, was not affected (KUMON et al. 1990).

Recent results showed that transgenic tobacco plants expressing the active-site mutant PAP were not resistant to virus infection, while transgenic tobacco expressing the C-terminal deletion mutant were resistant (TUMER et al. 1997). The ribosomes isolated from the C-terminal deletion mutant line were tested for depurination and no diagnostic rRNA fragment was released upon treatment with aniline. In contrast, ribosomes isolated from the transgenic line expressing PAP-v released the rRNA fragment upon treatment with aniline, indicating that ribosomes are depurinated in transgenic plants expressing PAP-v. The C-terminal deletion mutant PAP purified from transgenic tobacco plants did not deadenylate ribosomes in vitro (unpublished results). These results demonstrated that toxicity of PAP could be dissociated from its antiviral activity.

Analysis of transgenic plants indicated that expression of PAP-v induces synthesis of PR proteins and a very weak (less than two-fold) increase in salicylic acid levels (SMIRNOV et al. 1997). Reciprocal grafting experiments demonstrated that transgenic tobacco rootstocks expressing PAP-v induced resistance to tobacco mosaic virus infection in both *Nicotiana tabacum* NN and nn scions (SMIRNOV et al. 1997). Increased resistance to potato virus X (PVX) was also observed in *Nicotiana tabacum* nn scions grafted on transgenic rootstocks. PAP expression was not detected in wild-type scions or rootstocks that showed virus resistance, nor was there any increase in salicylic acid levels or PR protein synthesis. Grafting experiments with transgenic plants expressing an inactive PAP mutant demonstrated that an intact active-site of PAP is necessary for induction of virus resistance in wild-type scions. These results demonstrated that enzymatic activity of PAP is responsible for generating a signal that renders wild-type scions resistant to virus infection in the absence of increased salicylic acid levels and PR protein synthesis.

This mechanism is now extended by the observation that PAP can directly deadenylate viral RNA, *E. coli* rRNA and herring sperm DNA (BARBIERI et al. 1997). Incubations of PAP, PAPII and PAP-S in pmolar concentrations with DNA all resulted in the release of adenine in nmolar concentrations. These RIPs showed similar activity on *E. coli* rRNA; in contrast, their activity on bacteriophage MS2 RNA and tomato mosaic virus RNA was significantly less, the variability owing perhaps to the lack of required cofactors. However, this does not preclude their ability to deadenylate DNA and RNA substrates in vivo and increases the potential role of PAP as an antiviral agent, acting directly on the virus. These results suggest that RIPs inhibit viral infection by activating defense and stress-related signal transduction pathways. In addition, they could be affecting viral translation and replication by directly targeting viral nucleic acids in vivo. Further studies exploring the activity of PAP on viral nucleic acids and expression of cellular genes should lead to the elucidation of the mechanism of antiviral action.

6 Antifungal Activity and Possible Mechanisms

The rationale behind employment of ribosome inactivating proteins to achieve resistance against fungal pathogens is based on the ability of some RIPs to depurinate ribosomes of various fungi. Several successful examples of fungus resistance in transgenic plants expressing barley ribosome-inactivating proteins have been demonstrated. LOGEMANN et al. (1992) used barley endosperm RIP placed under the control of an inducible promoter and provided protection against fungal infection in transgenic tobacco plants. In another example, disease resistance in transgenic tobacco was shown to increase even further when constitutively expressed barley RIP was combined with barley class II chitinase, a gene with well documented lytic activity on fungal cell walls. JACH et al. (1995) showed that constitutive expression of barley endosperm RIP together with barley class-II chitinase yielded a significant increase in resistance to *Rhizoctonia solani* in transgenic tobacco. Products of both transgenes accumulated in the intercellular spaces of transgenic tobacco. It was postulated that the cytotoxic effect of the barley RIP on fungal cells was enhanced by chitinase action resulting in an increased amount of RIP entering fungal cells. MADDALONI et al. (1997) employed maize ribosome-inactivating protein b-32 under control of the potato wun1 gene promoter to increase the tolerance of transgenic tobacco to *Rhizoctonia solani*.

In contrast to abundant data concerning antiviral effects of PAP and models of PAP action, very little is known about other potential antimicrobial properties of PAP. Recent results showed that transgenic tobacco plants expressing PAP exhibited significant reduction in disease symptoms when inoculated with *Rhizoctonia solani*. Furthermore, resistance observed was not only due to the expression of the wild-type PAP transgene. Various nontoxic mutant forms of PAP also provided resistance against fungal infection, or caused a delay in disease progression (ZOU-BENKO et al. 1997).

The most resistant transgenic lines were those expressing the wild-type PAP cDNA and lines expressing the C-terminal truncated PAP. Both of these plant lines were previously shown to be resistant to plant viruses (LODGE et al. 1993; TUMER et al. 1997). Interestingly, although PAP has potent antiviral activity when exogenously applied, it does not inhibit fungal growth in plate assays in vitro (CHEN et al. 1991), suggesting that the fungal resistance observed in transgenic plants may be due to the action of PAP on the host plant. Indeed, analysis of transgenic lines showed elevated levels of pathogenesis-related proteins (PR proteins) known for their antifungal effect. Efficient constitutive expression of both isoforms, class I and class II, of chitinases and β-1,3-glucanases, and PR-1 was detected in transgenic lines expressing wild-type PAP, PAP-v, and the C-terminal deletion mutant. In contrast, induction of PR proteins by enzymatically inactive PAP-X was inefficient and less stable. Consequently, PAP-X expressing lines were the least resistant. Therefore both antifungal and antiviral activities of PAP seem to require an intact active-site. LIN et al. (1994) showed that tricholin, a ribosome inactivating protein from *Trichoderma viridae* has a complex effect on *Rhizoctonia solani*, demonstrated by decrease in amino acid uptake, inhibition of protein synthesis and cessation of growth. In the presence of tricholin, fungal cells exhibited altered polysome profiles attributed to the damage of ribosomal RNA.

In many cases salicylic acid plays a key role in local and systemic disease resistance, although it may be not the primary signal (VERNOOIJ et al. 1994). Our data suggest that PAP transgenic plants exhibit acquired resistance through a pathway which somehow circumvents salicylic acid, either by acting downstream or, more likely, involving an independent signal. Although PR protein expression was induced, increases in SA levels were not detected in transgenic plants expressing either the wild-type or the nontoxic mutant forms of PAP. Neither did these plants show HR-like lesions, which are characteristic of defense responses mediated by salicylic acid (ROSS 1961; RYALS et al. 1996). In addition, constitutive expression of basic chitinase, which normally is not induced during SAR, was observed in transgenic plants. Additional signaling pathways might be induced in PAP transgenic plants as was demonstrated in reciprocal grafting experiments (SMIRNOV et al. 1997). The absence of PR proteins in wild-type scions suggests that PAP induces a subset of host genes locally in cells expressing active forms of pokeweed antiviral protein.

Unlike virus resistance, antifungal action of PAP in transgenic plants may result from a dual effect, directly affecting pathogen cellular metabolism by inhibiting protein synthesis, and indirectly, via the action of host defense related genes which are constitutively upregulated.

7 Physiological Effects and Future Studies

Recent studies may help us define the function of RIPs in plants. RIPs may play a defensive role against plant pathogens. In addition, RIPs may provide a regulatory role in cell metabolism. It has been noted that RIP activity appears, or increases when seeds mature (FERRERAS et al. 1993) and in senescent or stressed leaves (STIRPE et al. 1996), in correlation with events leading to arrest of plant cell metabolism. Enzymatic activity on DNA may very well contribute to this regulation, because the deadenylation may be sufficient to disrupt transcription. Enzymatic activity on poly (A) tails and different RNA templates are suggestive of regulation of expression and cell metabolism.

PAP activity is influenced by the age of the plant and its environmental growing conditions. The translation inhibitory activity and the deadenylation of DNA are both significantly increased in the leaves of pokeweed during senescence or when subjected to heat or osmotic stress (STIRPE et al. 1996). Recent results indicate that human cells exposed to RIPs, such as PAP or ricin undergo rapid apoptosis (WADDICK et al. 1995). It is conceivable that the expression of PAP is increased in cells or tissues undergoing apoptosis during senescence or viral infection. If the action of PAP on RNA and on DNA observed in vitro were to occur in vivo, a direct effect of PAP on viral nucleic acids would lead to inhibition of viral infection. It is possible that PAP could kill cells by acting directly on DNA, providing an efficient mechanism for the induction of programmed cell death.

PAP is a very good candidate for biotechnological applications because of its broad specificity. The recent demonstration of the antifungal activity of PAP (ZOUBENKO et al. 1997), and the separation of its cytotoxicity from antiviral (TUMER et al. 1997) and antifungal activity (ZOUBENKO et al. 1997) suggest that PAP may be a useful agronomic tool to engineer broad spectrum disease resistance without any deleterious effects on transgenic plants. PAP mutants with reduced toxicity show promise not only in elucidating the mechanism of action of PAP, but also in biomedical and agricultural applications.

References

Aron GM and Irvin JD (1980) Inhibition of herpes simplex virus multiplication by the pokeweed antiviral protein. Antimicrob Agents Chemother 17:1032–1033

Balasundaram D, Dinman JD, Wickner RB, Tabor CW, Tabor H (1994) Spermidine deficiency increases +1 ribosomal frameshifting efficiency and inhibits Ty1 retrotransposition in Saccharomyces cerevisiae. Proc Natl Acad Sci USA 91:172–176

Barbieri I, Aron GM, Irvin JD, Stirpe F (1982) Purification and partial characterization of another form of the antiviral protein from the seeds of Phytolacca americana L (Pokeweed) Biochem J 203:55–59

Barbieri L, Bolognesi A, Cenini P, Falasca AI, Minghetti A, Garofano L, Guicciardi A, Lappi D, Miller SP, Stirpe F (1989) Ribosome-inactivating proteins from plant cells in culture. Biochem J 257:801–807

Barbieri L, Gorini P, Valbonesi P, Castiglioni P, Stirpe F (1994) Unexpected activity of saporins. Nature 372:624

Barbieri L, Valbonesi P, Bonora E, Gorini P, Bolognesi A, Stirpe F (1997) Polynucleotide:adenosine glycosidase activity of ribosome-inactivating proteins: effect on DNA, RNA and poly(A) Nucleic Acids Res 25:518–522

Benatti L, Nitti G, Solinas M, Valsasina B, Vitale A, Ceriotti A (1991) A saporin-6 cDNA containing a precursor sequence coding for a carboxyl-terminal extension. FEBS Lett 291:285–288

Bolognesi A, Barbieri L, Abbondanza A, Falasca AI, Carnicelli D, Battelli MG, Stirpe F (1990) Purification and properties of new ribosome inactivating proteins with RNA N-glycosidase activity. Biochim Biophys Acta 1087:293–302

Bonness MB, Ready MP, Irvin JD, Mabry TJ (1994) Pokeweed antiviral protein inactivates pokeweed ribosomes: implications for the antiviral mechanism. Plant J 5:173–183

Brigotti M, Carnicelli D, Alvergna P, Pallanca A, Sperti S, Montanaro L (1995) Differential up-regulation by tRNAs of ribosome-inactivating proteins. FEBS Lett 373:115–118

Carnicelli D, Brigotti M, Montanaro L, Sperti S (1992) Differential requirements of ATP and extraribosomal proteins for ribosome inactivation by eight RNA-N-glycosidases. Biochem Biophys Res Commun 182:579

Chaddock JA, Lord JM, Hartley MR, Roberts LM (1994) Pokeweed antiviral protein (PAP) mutations which permit E.coli growth do not eliminate catalytic activity towards prokaryotic ribosomes. Nucleic Acids Res 22:1536–1554

Chaddock JA, Monzingo AF, Robertus JD, Lord JM, Roberts LM (1996) Major structural differences between pokeweed antiviral protein and ricin A-chain do not account for their differing ribosome specificity. Eur J Biochem 235:159–166

Chen ZC, White RF, Antoniw JF, Lin Q (1991) Effect of pokeweed antiviral protein (PAP) on the infection of viruses. Plant Pathology 40:612–620

Chen ZC, Antoniw JF, White RF (1993) A possible mechanism for the antiviral activity of pokeweed antiviral protein. Physiol Mol Plant Pathol 42:249–258

Dinman JD, Icho T, Wickner RB (1991) A-1 ribosomal frameshift in a double-stranded RNA virus forms a Gag-Pol fusion protein. Proc Natl Acad Sci USA 88:174–178

Dinman JD (1995) Ribosomal frameshifting in yeast viruses. Yeast 11:1115–1127

Dinman JD, Wickner RB (1992) Ribosomal frameshifting efficiency and Gag/Gag-pol ratio are critical for yeast M1 double-stranded RNA virus propagation. J Virology 66:3669–3676

Duggar BM, Armstrong JK (1925) The effect of treating virus of tobacco mosaic with juice of various plants. Ann Mol Bot Gard 12:359

Endo Y, Tsurugi K (1987) RNA N-glycosidase activity of ricin A-chain: mechanism of action of the toxic lectin ricin on eukaryotic ribosomes. J Biol Chem 262:8128–8130

Endo Y, Tsurugi K, Lambert JM (1988) The site of action of six different ribosome-inactivating proteins from plants on eukaryotic ribosomes: the RNA N-glycosidase activity of the proteins. Biochem Biophys Res Commun 150:1032–1036

Erice A, Balfour HH Jr, Myers DE, Leske VL, Sannerud KJ, Kuebelbeck V, Irvin JD, Uckun FM (1993) Anti-human immunodeficiency virus type 1 activity of an anti-CD4 immunoconjugate containing pokeweed antiviral protein. Antimicrob Agents Chemother 37:835–838

Ferreras JM, Barbieri L, Girbes T, Battelli MG, Rojo MA, Arias FJ, Rocher MA, Soriano F, Mendez E, Stirpe F (1993) Distribution and properties of major ribosome-inactivating proteins (28S rRNA N-glycosidases) of the plant Saponaria officinalis L (Caryophyllaceae) Biochim Biophys Acta 1216:31–42

Frankel A, Schlossman D, Welsh P, Hertler A, Withers D, Johnston S (1989) Selection and character-ization of ricin toxin A-chain mutations in *Saccharomyces cerevisiae*. Mol Cell Biol 9:415–420

Funatsu G, Islam MR, Minami Y, Kung S, Kimura M (1991) Conserved amino acid residues in ribo-some-inactivating proteins from plants. Biochemie 73:1157–1161

Gessner SL, Irvin JD (1980) Inhibition of elongation factor 2-dependent translocation by pokeweed antiviral protein and ricin. J Biol Chem 255:3251–3253

Gesteland RF, Atkins JF (1996) Recoding: dynamic reprogramming of translation. Annu Rev Biochem 65:741–768

Gluck A, Endo Y, Wool IG (1992) Ribosomal RNA identity elements for ricin A-chain recognition and catalysis: analysis with tetraloop mutants. J Mol Biol 226:411–424

Grasso S, Jones P, White RF (1979) Inhibition of tobacco mosaic virus multiplication in tobacco pro-toplasts by the pokeweed inhibitor. Phytopathology 98:53–58

Hartley MR, Legname G, Osborn R, Chen Z, Lord MJ (1991) Single-chain ribosome inactivating pro-teins from plants depurinate *Escherichia coli* 23S ribosomal RNA. FEBS Lett 290:65–68

Hayashi S, Wu HC (1990) Lipoproteins in bacteria. J Bioenenerg Biomembr 22:451–471

Hong Y, Saunders K, Hartley MR, Stanley J (1996) Resistance to geminivirus infection by virus-induced expression of diathin in transgenic plants. Virology 220:119–127

Houston LL, Ramakrishnan S, Hermodson MA (1983) Seasonal variations in different forms of poke-weed antiviral protein, a potent inactivator of ribosomes. J Biol Chem 258:9601–9604

Huang PL, Chen HC, Kung HF, Huang PL, Huang P, Huang HI, Lee-Huang S (1992) Anti-HIV plant proteins catalyze topological changes of DNA into inactive forms. Biofactors 4:37–41

Hur Y, Hwang D-J, Zoubenko O, Coetzer C, Uckun F, Tumer NE (1995) Isolation and characterization of pokeweed antiviral protein in *Saccharomyces cerevisiae*: identification of residues important for toxicity. Proc Natl Acad Sci USA 92:8448–8452

Irvin JD (1975) Purification and partial characterization of the antiviral protein from *Phytolacca amer-icana* which inhibits eukaryotic protein synthesis. Arch Biochem Biophys 169:522–528

Irvin JD, Uckun FM (1992) Pokeweed antiviral protein: ribosome inactivation and therapeutic appli-cations. Pharmacol Ther 55:279–302

Irvin JD (1995) Antiviral proteins from *Phytolacca*. In: Chessin M, DeBorde D, Zipf A (eds) Antiviral proteins in higher plants. CRC, Boca Raton, pp 65–94

Irvin JD, Kelly T, Robertus JD (1980) Purification and properties of a second antiviral protein from *Phytolacca americana* which inactivates eukaryotic ribosomes. Arch Biochem Biophys 200:418–425

Islam MR, Kung S, Kimura Y, Funatsu G (1991) N-acetyl-D-glucosamine asparagine structure in ri-bosome-inactivating proteins from the seeds of *Luffa cylindrica* and *Phytolacca americana*. Agricult Biol Chem 55:1375–1381

Jach G, Gornhardt B, Mundy J, Logemann J, Pinsdorf E, Leah R, Schell J, Maas C (1995) Enhanced quantitative resistance against fungal disease by combinatorial expression of different barley anti-fungal proteins in transgenic tobacco. Plant J 8:97–109

Kataoka J, Habuka N, Furuno M, Miyano M, Takanami Y, Kiowai A (1991) DNA sequence of *Mirabilis* antiviral protein (MAP), a ribosome-inactivating protein with an antiviral property, from *Mirabilis jalapa* L. and its expression in *Escherichia coli*. J Biol Chem 266:8426–8430

Kumon K, Sasaki J, Sejima M, Takeuchi Y, Hayashi Y (1990) Interactions between tobacco mosaic virus, pokeweed antiviral proteins, and tobacco cell wall. Phytopathology 80:636–641

Kung S, Kimura M, Funatsu G (1990) The complete amino acid sequence of antiviral protein from the seeds of pokeweed (*Phytolacca americana*) Agricult Biol Chem 34:3301

Lam YH, Wong B, Wang RN-S, Yeung HW, Shaw PC (1996) Use of trichosanthin to reduce infection by turnip mosaic virus. Plant Sci 114:111–117

Lee-Huang S, Kung HF, Huang PL, Bourinbaiar AR, Morell JL, Brown JH, Huang PL, Tsai WP, Chen AY, Huang HI (1994) Human immunodeficiency virus type 1 (HIV-1) inhibition, DNA-binding, RNA-binding, and ribosome inactivating activities in the N-terminal segments of the plant anti-HIV protein GAP31. Proc Natl Acad Sci USA 91:12208–12212

Legname G, Bellosta P, Gromo G, Modena D, Keen JN, Roberts IN, Lord JM (1991) Nucleotide sequence of cDNA coding for dianthin 30, a ribosome inactivating protein from *Dianthus car-yophyllus*. Biochim Biophys Acta 1090:119–122

Li MX, Yeung HW, Pan LP, Chan SI (1991) Trichosanthin, a potent HIV-1 inhibitor, can cleave supercoiled RNA in vitro. Nucleic Acids Res 19:6309–6312

Lin A, Lee TM, Rern JC (1994) Tricholin, a new antifungal agent from *Trichoderma viride*, and its action in biological control of *Rhizoctonia solani*. J Antibiotics 47:799–805

Lin Q, Chen ZC, Antoniw JF, White RF (1991) Isolation and characterization of a cDNA clone encoding the anti-viral protein from *Phytolacca americana*. Plant Mol Biol 17:609–614

Ling J, Liu W, Wang TP (1994) Cleavage of supercoiled double-stranded RDNA by several ribosome-inactivating proteins in vitro. FEBS Lett 345:143–146

Lodge JK, Kaniewski WK, Tumer NE (1993) Broad-spectrum virus resistance in transgenic plants expressing pokeweed antiviral protein. Proc Natl Acad Sci USA 90:7089–7093

Logemann J, Jach G, Tommerup H, Mundy J, Schell J (1992) Expression of barley ribosome-inactivating protein leads to increased fungal protection in transgenic tobacco plants. Biotechnology 10:305–308

Lord JM, Roberts LM, Robertus JD (1994) Ricin: structure, mode of action and some current applications. FASEB J 8:201–208

Maddaloni M, Forlani F, Balmas V, Donini G, Stasse L, Corazza L, Motto M (1997) Tolerance to the fungal pathogen *Rhizoctonia solani* AG4 of transgenic tobacco expressing the maize ribosome-inactivating protein b-32. Transgenic Res 6:393–402

Marchant A, Hartley MR (1995) The action of pokeweed antiviral protein and ricin A-chain on mutants in the α-sarcin loop of *Escherichia coli* 23S ribosomal RNA. J Mol Biol 254:848–855

Menninger JR (1975) Peptidyl transfer RNA dissociates during protein synthesis from ribosomes of Escherichia coli. J Biol Chem 251:3392–3398

Mock JWY, Ng TB, Wong RNS, Yao QZ, Yeung HW, Fong WP (1996) Demonstration of ribonuclease activity in the plant ribosome-inactivating proteins alpha- and beta-momorcharins. Life Sci 59:1853–1859

Monzingo AF, Robertus JD (1992) X-ray analysis of substrate analogs in the ricin A chain active site. J Mol Biol 227:1136–1145

Monzingo AF, Collins EJ, Ernst SR, Irvin JD, Robertus JD (1993) The 2.5 A structure of pokeweed antiviral protein. J Mol Biol 233:705–715

Myers DE, Irvin JD, Smith RS, Kuebelbeck VM, Tuel-Ahlgren L, Uckun FM (1991) Large scale production of a pokeweed antiviral protein (PAP) containing immunotoxin B43-PAP directed against the CD19 human B-lineage differentiation antigen. J Immunol Methods 136:221

Nicolas E, Beggs JM, Haltiwanger BM, Taraschi TF (1997) Direct evidence for the deoxyribonuclease activity of the plant ribosome inactivating protein gelonin. FEBS Lett 406:162–164

Nicholls A, Sharp K, Honig B (1991) Protein folding and association: insights from the interfacial and thermodynamic properties of hydrocarbons. Prot Struct Funct Genet 11:281–296

Obrig TG, Moran TP, Colinas RJ (1985) Ribonuclease activity associated with the 60 S ribosome-inactivating proteins ricin A, phytolaccin and shiga toxin. Biochem Biophys Res 130:879–884

Poyet J-L, Radom J, Hoeveler A (1994) Isolation and characterization of a cDNA clone encoding the pokeweed antiviral protein II from *Phytolacca americana* and its expression in *E. coli*. FEBS Lett 347:268–272

Poyet J-L, Hoeveler A (1997) cDNA cloning and expression of pokeweed antiviral protein from seeds in *Escherichia coli* and its inhibition of protein synthesis in vitro. FEBS Lett 406:97

Ready MP, Brown DT, Robertus JD (1986) Extracellular localization of pokeweed antiviral protein. Proc Natl Acad Sci USA 84:5053–5056

Ready M, Kim Y, Robertus JD (1991) Site-directed mutagenesis of ricin a-chain and implications for the mechanism of action. Proteins 10:270–278

Robertus J (1991) The structure and action of ricin, a cytotoxic N-glycosidase. Semin Cell Biol 2:23–30

Rodes III TL, Irvin JD (1981) Reversal of the inhibitory effects of the pokeweed antiviral protein upon protein synthesis. Biochim Biophys Acta 652:160–167

Ross AF (1961) Systemic acquired resistance induced by localized virus infections in plants. Virology 14:340–358

Ryals JA, Neuenschwander UH, Willits MG, Molina A, Steiner H-Y, Hunt M (1996) Systemic acquired resistance. Plant Cell 8:1809–1819

Shepherd RJ, Fulton JP, Wakeman RJ (1969) Properties of a virus causing pokeweed mosaic. Phytopathology 59:219–222

Smirnov S, Shulev V, Tumer NE (1997) Expression of pokeweed antiviral protein in transgenic plants induces virus resistance in grafted wild-type plants independently of salicylic acid accumulation and pathogenesis-related protein synthesis. Plant Physiol 114:1113–1121

Stevens WA (1981) Effect of inhibitors of protein synthesis from plants on tobacco mosaic virus infection. Experientia 37:257–259

Stirpe F, Barbieri L, Gorini P, Valbonesi P, Bolognesi A, Polito L (1996) Activities associated with the presence of ribosome-inactivating proteins increase in senescent and stressed leaves. FEBS Lett 382:309–312

Taylor S, Massiach A, Lomonossoff G, Roberts LM, Lord JM, Hartley M (1994) Correlation between the activities of five ribosome-inactivating proteins in depurination of tobacco ribosomes and inhibition of tobacco mosaic virus infection. Plant J 5:827–835

Teltow GJ, Irvin JD, Aron GM (1983) Inhibition of Herpes Simplex Virus DNA Synthesis by Pokeweed Antiviral Protein. Antimicro Agents Chemo 23:390–396

Thompson RC, Dix DB (1982) Accuracy of protein biosynthesis. J Biol Chem 257:6677–6682

Tomlinson JA, Walker VM, Flewett TH, Barclay GR (1974) The inhibition of infection by cucumber mosaic virus and influenza virus by extracts from Phytolacca americana. J Gen Virol 22:225–232

Tumer NE, Hwang D-J, Bonness M (1997) C-terminal deletion mutant of pokeweed antiviral protein inhibits viral infection but does not depurinate host ribosomes. Proc Natl Acad Sci USA 94:3866–3871

Tumer NE, Parikh BA, Li P, Dinman JD (1998) The pokeweed antiviral protein specifically inhibits Ty1-directed + 1 ribosomal frameshifting and retrotransposition in *Saccharomyces cerevisiae*. J Virology 72:1036–1042

Uhlenbeck OC (1995) Keeping RNA happy. RNA 1:4–6

Ussery MA, Hardesty B (1977) Inhibition of poliovirus replication by a plant antiviral peptide. Ann NY Acad Sci 284:431–440

Vernooij B, Friedrich L, Morse A, Reist R, Kolditz-Jawhar R, Ward E (1994) Salicylic acid is not the translocated signal responsible for inducing systemic acquired resistance but is required in signal transduction. Plant Cell 6:959–965

Waddick KG, Gunther R, Chelstrom LM, Chandan-Langlie M, Myers DE, Irvin JD (1995) In vitro and in vivo anti-leukemic activity of B43 (anti-CD-19)-pokeweed antiviral protein immunotoxin against radiation resistant human pre-B acute lymphoblastic leukemia cells. Blood 86:4228–4233

Wales R, Roberts LM, Lord MJ (1993) Addition of an endoplasmic reticulum retrieval sequence to ricin A-chain significantly increases its cytotoxicity to mammalian cells. J Biol Chem 268:23986–23990

Zarling JM, Moran PA, Haffar O, Sias J, Richman DD, Spina CA, Myers DE, Kuebelbeck V, Ledbetter JA, Uckun FM (1990) Inhibition of HIV replication by pokeweed antiviral protein targeted to CD4 + cells by monoclonal antibodies. Nature 347:92–95

Zoubenko O, Uckun F, Hur Y, Chet I, Tumer N (1997) Plant resistance to fungal infection induced by nontoxic pokeweed antiviral protein mutants. Nature/Biotech 15:992–996

Transgenic Plants as Edible Vaccines

L. RICHTER and P.B. KIPP

1 Introduction

Vaccinations are among the more cost-effective health care procedures (HAUS-DORFF 1996). In the United States and Europe, a majority of newborn children are vaccinated against ten diseases (Recommended childhood immunization schedule from Center for Disease Control, USA). Recently in the United States hepatitis B was added to the recommended vaccinations for infants. Of these different immunizations, only one is an oral vaccine with the others requiring injections. In addition, approximately thirty new vaccines are currently under development in the United States (DIVISION OF MICROBIOLOGY AND INFECTIONS DISEASES 1995; 1996).

Boyce Thompson Institute for Plant Research Inc., Tower Rd., Ithaca, NY 14850, USA

In contrast to Western countries, vaccines are in limited use in many developing countries, with children receiving the "Expanded Program on Immunization" (EPI) vaccines against six diseases. These are recommended, and in many countries financed, by the World Health Organization. The cost of vaccines is one factor preventing further use of vaccination, leaving hundreds of thousands of children susceptible to preventable diseases (HAUSDORFF 1996). The principle costs of most commercial vaccines are production, packaging and delivery. Injectable vaccines incur further expenses related to the use and disposal of needles and syringes, trained personnel to administer injections, and refrigeration required during shipping and storage. These same economic factors prevent widespread vaccination of livestock, poultry and swine against preventable diseases.

The Children's Vaccine Initiative (MITCHELL et al. 1993) called for new technologies to make vaccines more widely available. This includes low cost production systems and to further develop oral vaccines. Oral vaccines are desirable due to their ease of administration and patients acceptance of noninjected vaccinations. Another advantage is that oral vaccines may stimulate production of mucosal antibodies more effectively than injected vaccines. This is important as the mucosal immune system is a first line of defense against many disease organisms.

Vaccines are designed to elicit an immune response without causing disease. Typical vaccines are composed of killed or attenuated disease-causing organisms. Recombinant subunit vaccines are desirable as an alternative with potentially fewer side effects than delivering the whole organism. Recombinant subunit vaccines do not contain an infectious agent, and thus are safer to administer and prepare, and doses are more uniform. Advances in molecular biology of diseases have identified many candidate proteins or peptides that may function as effective subunit vaccines. Recombinant vaccines have potential for being highly effective in preventing disease, both in humans and animals, but are rather costly to produce, and therefore are in limited use worldwide. The choice of which system to use to produce a recombinant vaccine must take into consideration their advantages and disadvantages, costs of production, and the amount of product required on a global scale. For some vaccine antigens, transgenic plants may provide an ideal expression system, in which transgenic plant material can be fed directly to subjects as their oral dose of recombinant vaccine.

2 Transgenic Plants as Recombinant Protein Production Systems

A variety of production systems are available for recombinant proteins including; yeast, insect cell culture, mammalian cell culture, bacteria, transgenic animals, and plants. Currently, the most common large scale production systems for proteins are genetically engineered bacteria and yeast, due to these organisms relative ease of manipulation and rapid predictable growth. Recombinant proteins overexpressed in genetically engineered bacteria and yeast are extensively purified to remove host proteins and compounds. These processes add to the cost of recombinant proteins.

Transgenic plants provide an alternative system that can be scaled up to high production capacity. A great practical advantage of producing vaccines in transgenic plants is the ability to directly use edible plant tissues for oral administration without purification. If a "food source" plant is used for production of edible vaccines, some of the problems associated with other large scale systems during purification need not be considered, such as host toxins or additives during purification. If processing of the "edible vaccine" is performed, this would be simpler than purification and might include sterilization and packaging much as baby food containing banana or other raw foods. This article does not address bioproduction in plants that requires purification of the product.

The development and refinement of plant genetic engineering techniques and an improving knowledge of plant molecular biology is continuously expanding the potential of plant biotechnology. Transgenic plants show a promising capacity to express heterologous proteins. A great deal of research is focused toward understanding the fundamental processes of transgene expression and recombinant protein accumulation, stability, and processing in plants.

2.1 Choice of Plant Species for Recombinant Vaccine Production

The choice of which plant species to use for production and delivery of an oral vaccine can readily be tailored to the subject to be vaccinated. The food plants for which transformations have been reported are: alfalfa, apple, asparagus, banana, barley, cabbage, canola/rape seed, cantaloupe, carrot, cauliflower, cranberry, cucumber, eggplant, flax, grape, kiwi, lettuce, lupins, maize, melon, papaya, pea, peanut, pepper, plum, potato, raspberry, rice, service berry, soybean, squash, strawberry, sugar beet, sugarcane, sunflower, sweet potato, tomato, walnut, and wheat (JAMES and KRATTIGER 1996; MAY et al. 1995). For many farm animals, maize may be an appropriate crop for vaccine production and delivery, as corn is a major component of many animal feeds. Younger animals, that cannot eat solid food, may be immunized passively by suckling milk containing maternal antibodies provided by vaccinated mothers (FERNANDEZ et al. 1996). For human vaccination, a crop such as banana is desirable. Bananas are readily eaten by most babies, are consumed uncooked (to prevent antigen protein denaturation), and are indigenous to many developing countries (decreasing transportation costs) where low cost vaccines are needed most.

2.2 Optimization of Recombinant Protein Production

Transgenic plants are commonly produced by two methods, *Agrobacterium*-mediated transformation or particle bombardment. With either method, the transDNA can be inserted into almost anywhere in the plant genome. The position of insertion usually has an effect on the expression of the transgenes, therefore several independent transformed plant lines must be analyzed for their expression level to find the best expressing lines.

To economically utilize plants to produce proteins, it is crucial to maximize transgene expression and recombinant protein accumulation. Technical difficulties can arise when employing transgenic plants to produce foreign proteins, often resulting in a low yield of recombinant protein. The reported yields of recombinant proteins from transgenic plants are highly variable. Many recombinant proteins comprise only about 0.01% of the total protein, although exceptions are common. Recombinant avidin expressed in transgenic maize seed accumulated to a level of approximately 2% of the total soluble extractable protein (Hood et al. 1997). The factors permitting high levels of accumulation are not clear, and transgenes must be assessed on a case by case basis.

Yields of recombinant proteins can be increased by augmenting protein stability, mRNA stability and/or translatability, and enhancing promoter strength. For some recombinant proteins, accumulation can be increased by targeting the recombinant protein to a subcellular compartment, such as the endoplasmic reticulum or vacuole (see Sect. 3.2.1). Specific sequences within an mRNA can affect the stability and translatability of the message in plant cells. Such sequences include mRNA destabilizing motifs, cryptic splice and polyadenylation signals, (A/U rich regions) as well as suboptimal codon usage (Newman et al. 1993; Ohme-Takagi et al. 1993; Taylor and Green 1995). Several examples of aberrant splicing of heterologous cDNAs in plant hosts have been published, most notably for the green fluorescent protein (Rouwendal et al. 1997). By creating a synthetic gene, many (if not all) of these potentially troublesome sequences can be removed. In several cases, the use of "plant-optimized" synthetic genes enhanced the accumulation of recombinant protein dramatically (Perlak et al. 1991; Rouwendal et al. 1997). In addition, the use of a strong promoter is critical to maximizing recombinant protein yields. It may be desirable to employ synthetically inducible or tissue specific promoters as described below for LT-B.

2.3 Posttranslational Modifications of Recombinant Proteins

Some eukaryotic proteins require extensive posttranslational modifications to assume active conformations. This is a major limitation of prokaryotic expression systems. Plants utilize chaperones and protein disulfide isomerase to facilitate protein folding, and to prevent large and complex proteins aggregates from forming protein aggregates (Sturm 1995). Plants also glycosylate proteins. An important issue in the expression of vaccine antigens is glycosylation as carbohydrate groups contribute to immunogenicity. The capacity of plants to glycosylate heterologous proteins will need to be determined for each antigen.

2.4 Model Plant Systems

Typical experiments investigating in planta recombinant protein expression employ plant model systems such as; tobacco, potato, tomato and maize (Ball et al. 1996; Haq et al. 1995; Ma et al. 1997; Mason et al. 1992; McGarvey et al. 1995). Each of these species is readily transformed, and can be propagated and regenerated in tissue culture by established protocols. Historically, transgenic work has been carried out in tobacco due to its ease of transformation, rapid growth, and robust regeneration. Transgenic tobacco engineered to express a recombinant vaccine protein is not an ideal material for oral delivery due to high levels of alkaloids and nicotine, although low alkaloid varieties exist that can be used in certain applications (MA et al. 1997). Potato, tomato and maize have both practical and molecular biological advantages as oral vaccine production systems over tobacco. Potato, tomato, and maize each produce edible plant tissue, which can be fed to animal subjects uncooked. A fundamental understanding of gene expression leading to tuber, tomato, and ear development permits the construction of tissue specific recombinant protein expression vectors utilizing regulatory elements from tissue specific genes, such as the patatin promoter of potato (WENZLER et al. 1989). A strategy employing tissue specific expression is designed to accumulate recombinant protein only in the tissue to be fed as a vaccine. Confining expression of an heterologous protein to a specific organ of the plant can be beneficial to the regeneration of transgenic plants, as high levels of recombinant protein accumulation can be harmful to plant development (NAWRATH et al. 1994) (see Sect. 3.2.1).

2.4.1 Potato

Our research group has extensively used potato for an experimental vaccine production and delivery system. Advantages of potato include: (a) efficiency of genetic transformation by *Agrobacterium tumefaciens* with relatively short transformation and generation times (DE BLOCK 1988), (b) clonal propagation, (c) availability of tissue specific promoters and capability of microtuber production, (d) storage potential, and (e) the ability to feed raw tubers to test animals and human volunteers.

Clonal propagation allows stable production when scaling up a transgenic plant line. In instances where constitutive promoters are employed, regenerating tissue can be screened for expression of transgenes directly. However, when tissue specific promoters are employed, the screening process is typically delayed, as development of specific organs does not occur until the plants are more mature. Putative potato transgenics can be forced to prematurely produce "microtubers," which are small but morphologically similar to naturally produced tubers (PERL 1991). Thus, potato transgenics can be readily screened for tuber specific expression by microtuberization while in tissue culture.

Tubers are the storage organs of the potato plant and can be stored for a period of time before being consumed. Foreign proteins may be able to survive for long periods of time in a tuber as long as the tuber does not sprout or become damaged. Each recombinant protein would have to be tested for its longevity within the tuber tissue. The benefits of being able to store tubers without any processing may make potatoes very desirable as a vaccine production system for those animals that do eat raw potatoes.

Potato tubers can be fed to mice and human volunteers for immunogenicity studies. The first experiments testing whether "edible vaccines" would promote an immune response were done by feeding raw potatoes to mice and are described in later sections for two candidate vaccines, LT-B and NVCP. The first human clinical trial testing "edible vaccines" also utilized raw tubers and is described below (Sect. 3.2.1).

One disadvantage of potato tubers is that they are not high in protein content and thus the level of recombinant protein must be higher on a percentage total protein basis than tissues containing more protein. To explain further, potatoes are approximately 2% protein by weight so 1 g tuber tissue contains at most 20 mg protein. For 1 g tuber to contain 1 mg recombinant protein, the recombinant protein must be 5% of the total protein in the tuber tissue. For tissues with a higher protein content, such as 10% protein, to produce 1 mg recombinant protein per gram tissue requires an expression level of foreign protein of only 1% of the total protein.

2.4.2 Tomato

The advantages of tomatoes include (a) ease of transformation (Mc Cormick 1991), (b) capability of genetic crossing to combine transgenes for production of multiple subunit vaccines into one plant, (c) production of seeds that can be stored long-term, and (d) tomatoes are eaten raw by humans. The disadvantages include the necessity of creating homozygotes for continuation of the line or analysis of every individual offspring for its transgene content and also, tomatoes, as with potatoes, are not high in protein content.

2.4.3 Banana

Transgenic bananas are our choice of a plant for production of oral vaccines for human use. Advantages of bananas include (a) ease of feeding, (b) consumed uncooked, (c) are grown in many developing countries where clonal propagation technology is already in use. The current disadvantages of bananas are in the technical difficulty in creating transgenic banana plants. Regeneration and growth to maturity and fruit production can take up to 3 years. This time period necessitates the use of model plants to determine the best parameters for high-level expression of a vaccine antigen in plant tissue before initiating work in banana plants. The protein content of banana fruit is about 1% (Stover and Simmonds 1987). Recent studies in molecular biology of bananas have identified several genes that are upregulated during ripening of banana fruit (Clendennen and May 1997) whose promoters may be appropriate for tissue specific expression of edible vaccines.

3 Candidate Vaccine

Typical vaccines have been composed of killed disease organisms (Salk polio vaccine), attenuated live organisms (Sabin polio vaccine) or strains of a disease organism whose host differs from the species being vaccinated (Jenner's cowpox). Research in immunology of several diseases has led to the proposal of subunit vaccines as an alternative to these historical approaches. A subunit vaccine can be composed of one or more proteins or components of the disease organism that induce the host to mount a protective immune response. An anti-idiotope may be considered a subunit vaccine if it promotes the desired immune response (GREEN-SPAN and BONA 1993). An example of a subunit vaccine is the hepatitis B surface antigen described below.

The hepatitis B surface antigen (HBsAg), has successfully been used as a vaccine against hepatitis B virus (HBV). HBsAg is the viral capsid protein and occurs in three forms in the native viral capsids differing in their amino ends. The commercial subunit vaccines marketed by Merck and SmithKline Beecham are composed of the shortest form, the S protein. The vaccine is made in a yeast production system and assembles into spherical particles or virus-like particles (VLPs) very similar to native HBV capsids. This was the first recombinant subunit vaccine. The vaccine is administered in three injections containing 10–20µg HBsAg in each dose. Figure 1 shows VLPs composed of recombinant HBsAg produced in transgenic plant tissue.

VLPs have excellent potential as effective vaccines. VLPs form by self-assembly of viral surface proteins. The intact particles appear very similar to the true virus, the VLPs therefore often present antigens that the immune system responds to as it would an inert or killed, fixed virus. Production of recombinant VLPs in the baculovirus insect cell culture system has been described for the following disease agents:

– African horse sickness virus (ROY et al. 1996)
– Bluetongue virus (ROY et al. 1994)
– Dengue virus (SUGRE et al. 1997)
– Hv190SV (HUANG et al. 1997)
– Hawaii virus (GREEN et al. 1997)
– Hepatitis B (KOLETZKI et al. 1997; MASON et al. 1992)

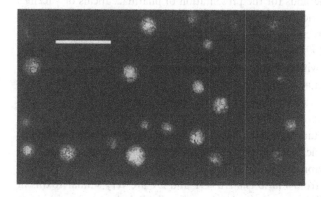

Fig. 1. Electron micrograph of HBsAg VLPs isolated from tobacco leaf extracts. *Bar.* 100 nm. (EM courtesy of Hugh S. Mason)

- Hepatitis E (TSAREV et al. 1993)
- Norwalk virus (BALL et al. 1996)
- Papillomaviruses (KIRNBAUER et al. 1996)
- Parvovirus (SEDLIK et al. 1997)
- Rotavirus (CRAWFORD et al. 1994)
- Rabbit hemorrhagic virus (LAURENT et al. 1994)
- Rhubella (GRANGEOT-KEROS and ENDER 1997)

Many of these might be candidates for production in plants as subunit vaccines. If assembly of the VLPs can be accomplished in an *in vitro* system the potential is available for assembly within a plant cell. Two demonstrations of VLP formation in plants are described below, Norwalk virus capsid protein (NVCP) and hepatitis B surface antigen (HBsAg).

3.1 Oral Immunogenicity and Adjuvants

A successful oral vaccine stimulates the mucosal immune system to produce a protective immune response against the disease agent. In general, mucosal IgA is required as well as serum IgG. The study of mucosal immune response has recently blossomed with new insights being gained on how to stimulate this part of the immune system. Stimulating the immune system through the mucosa to produce a similar level of circulating antibodies as compared to parenteral injection usually requires larger (100 times more) and/or multiple doses of antigen than is needed for parenteral delivery (CZERKINSKY and HOLMGREN 1995). An oral vaccine must survive the proteases and low pH of the stomach to reach the effector sites in the intestines where constant sampling of the milieu occurs by the M cells. The M cells transport particles and peptides to lymphoid tissue underlying the mucous membrane. Particles seem to be taken up more effectively by M cells than soluble peptides which are much less efficient immunogens. Antigens that form multimeric or heteromeric structures are likely to be better candidates for mucosal vaccines than soluble proteins.

There are several methods for the presentation of pharmaceuticals or vaccines by the oral route; enteric coated tablets, encapsulation in lipid microspheres, phospholipid-calcium precipitates, and mucoadhesive polymers (DANHEISER 1997; KUNTZ and SALTZMAN 1997). Production of oral vaccines in plants provides a built in encapsulation system with cell walls and cell membranes protecting the candidate vaccine through the stomach. Digestion of the plant material may provide a slow release of the candidate vaccine throughout the digestive system allowing presentation to the mucosal immune system.

Adjuvants are substances that enhance an immune response, and are commonly used with injectable vaccines. Several adjuvants are under study for mucosal use, including the heat-labile enterotoxin (LT) from enterotoxigenic *E. coli* (ETEC) and cholera toxin (CT) from *Vibrio cholera*. CT and LT are very similar toxins in structure, function, and immunological properties. Each holotoxin molecule is composed of one A subunit and five B subunits that form a pentamer. The B subunits, when assembled, can bind to the GM_1 gangliosides on the surface of the M cells within the digestive tract. For animal studies CT or LT can be added to a substance given orally to enhance the immune response. The toxic affects of CT and LT prevent their use in humans, however, several mutants of CT and LT are under study for potential use as mucosal adjuvants (DICKINSON and CLEMENTS 1996).

It is desirable to use plant-based edible vaccines without an adjuvant; however, for some antigens it may be necessary. If this is the case it is possible to engineer the vaccine producing plants to also make the adjuvant (see Sect. 3.2.1) or a mutant form of adjuvant that would not produce toxic affects. This would allow coad-ministration of the antigen with the adjuvant. Usually the amount of adjuvant is much smaller than the amount of antigen when presented orally. For example studies described below for oral immunization with NVCP used 40–80µg NVCP plus 10µg CT (BALL et al. 1996; MASON et al. 1996).

3.2 Vaccines Against Bacterial Pathogens

The first description of an antigen being produced in plants was in a European patent (CURTISS and CARDINEAU 1989). Curtiss and Cardineau engineered tobacco plants to produce the surface protein antigen A (spaA) from *Streptococcus mutans*. The authors calculated that the recombinant spaA protein represented 0.02% of the total protein in leaf samples. Data from oral immunogenicity studies were reported in which purified recombinant spaA protein made in *E. coli* was mixed with mouse chow and fed to mice. The mice developed sIgA specific for spaA in their saliva. Although a study using lyophilized tobacco extracts containing spaA was described in which the plant material was added to mouse chow, no data was reported so this may have been a theoretical study for purposes of the patent application. No further reports on these studies have been published.

3.2.1 LT-B as a Vaccine Against Cholera and ETEC

The significant events in development of an edible vaccine against cholera and ETEC are:

– 1992: create constructs for producing LT-B in transgenic plants
– 1993: produce recombinant LT-B protein in tobacco plants
– 1993: characterize transgenic product: antigenicity and pentamer assembly
– 1993: expression of LT-B in potato plants
– 1994: animal studies with purified antigen
– 1994: direct feeding studies with mice eating transgenic tubers producing LT-B
– 1995: creation of synthetic LT-B gene to enhance antigen production
– 1996: animal feeding studies with high-level LT-B expressing tubers
– 1997, July: FDA approval for human trials
– 1997, Oct.: human phase I clinical trials in United States

In 1995 Haq et al. published the first "proof of concept" for edible vaccines by demonstrating that mice fed doses of transgenic tubers producing the B subunit of the LT toxin (LT-B) produced both serum and mucosal antibodies against the LT-B protein. Both LT and CT toxins cause a cascade of events within the epithelial cells of the intestines resulting in diarrhea. It has been shown that these toxins cross react immunologically such that antibodies against one can block the effect of either toxin, and therefore immunization against CT or LT gives some protection against both disease agents. Both of these toxins are very immunogenic and can be used as adjuvants as discussed above (see Sect. 3.1).

Haq et al. (1995) showed that recombinant LT-B protein can be synthesized in tobacco and potato plants and that the LT-B maintained its ability to self-assemble into pentameric structures. The plant produced LT-B bound GM_1 gangliosides and was recognized by anti-LT antibodies. The tobacco produced LT-B was crudely purified and used for initial studies of oral immunization of mice. Subsequently, mice were fed raw tuber material expressing LT-B. Ingestion of the LT-B tubers induced anti-LT-B antibodies in the mice, both mucosal sIgA and serum IgG. As cholera and ETEC do not cause diarrhea in mice, these mice could not be challenged to determine if their immune response was protective. However, an *in vitro* assay using Y-1 adrenal cells showed that the antibodies could neutralize the LT toxin. Further *in vivo* experiments using a patent (intact) mouse model have shown that mice fed transgenic tubers expressing LT-B were partially protected from the toxic effects of the LT toxin (MASON et al. 1998). These results are encouraging since the tubers used for these experiments did not produce large amounts of antigen.

First attempts to produce LT-B in tobacco resulted in very low levels of expression, less than 0.01% total soluble protein. The level was increased by engineering the recombinant protein to include an endoplasmic reticulum retention signal (amino acids SEKDEL) at its carboxyl terminus. This modification resulted in a threefold increase over initial expression without the SEKDEL. The level of recombinant protein produced in the transgenic tubers was 3–4µg LT-B per gram fresh tuber weight.

Recent work by T. Haq and H. Mason to increase expression levels of recombinant LT-B by designing and creating a "plant-optimized" gene encoding LT-B has successfully resulted in potato tubers expressing up to 10–20µg per gram fresh weight (MASON et al. 1996 and 1998). The synthetic LT-B gene was designed with a consideration of both dicot and monocot plant preferences for codons, A-T richness, putative polyadenylation and cryptic splicing signals, and potential RNA destabilizing sequences. One amino acid change was introduced into the LT-B protein in the signal sequence for ease of cloning. This should be cleaved off in the processing of the LT-B protein.

In potato leaves, accumulation of the synthetic gene-encoded LT-B protein was as high as 1.8% of the total soluble protein, a 10-fold increase compared to the wild-type coding sequence. In tissue culture microtubers, the synthetic gene expressed recombinant LT-B up to 0.5% of the total soluble protein, a level 50-fold higher than the native LT-B gene with the SEKDEL endoplasmic retention signal. The levels of expression from the synthetic gene in tubers were 125-fold higher than the original tuber-specific construct using the native gene. Several of the better lines were grown into mature plants in greenhouses or light rooms. A consequence of the high constitutive LT-B protein production in the plants containing the synthetic gene was that the plants were stunted compared to wild-type plants. This complication was overcome by using the patatin promoter, a tuber specific promoter, driving expression of the synthetic LT-B gene. Transformation of potato plants with this construct resulted in healthy looking plants and tubers producing 10--20 μg recombinant LT-B per gram fresh tuber. With this level of expression, it is feasible to consider human efficacy trials.

Previous human trials with the B subunit of cholera toxin used up to 1 mg CT-B per oral dose (CLEMENTS et al. 1990; SCERPELLA et al. 1995). To use the same amount of plant produced LT-B, volunteers ingested 50–100g raw tuber per dose in phase I trials started in October 1997 at the University of Baltimore (TACKET et al. 1998). This was the first demonstration that recombinant antigens delivered orally in raw plant material without adjuvant elicit immune responses in human subjects. The volunteers had no food or water for 90 min before and after their edible vaccine dose and the raw potatoes were peeled and cubed immediately before ingestion. At least fourfold increases in serum IgG anti-LT, indicative of a systemic immune response, were measured in ten of eleven volunteers who ate the LT-B producing tubers. Furthermore, five out of ten volunteers developed at least a fourfold increase in sIgA measured from stool samples, indicating a mucosal immune response. These immune responses are comparable to immune responses observed in previous studies in which volunteers were challenged with 10^9 virulent enterotoxigenic E. coli after drinking sodium bicarbonate. These studies establish a fundamental proof of concept for transgenic plant oral vaccines.

3.3 Vaccines Against Viral Pathogens

3.3.1 Hepatitis B

Hepatitis B virus (HBV) infects over 300 million people worldwide and "kills about one million people each year, but it is preventable by vaccine" (WORLD HEALTH ORGANIZATION 1995). As described above (see Sect. 3), the commercial vaccines against HBV include use of vaccines composed of the recombinant protein hepatitis B surface antigen (HBsAg). Since the commercial HBV vaccines are effective and large quantities of antigen are expensive, there has been little incentive until now to determine whether HBsAg will be orally immunogenic. Recombinant HBsAg was the first recombinant vaccine and it is made with only one transgene. It also has been shown to be sufficient for protection as a parenteral vaccine and therefore, it was a model for a first attempt at production of a plant produced vaccine. HBsAg was expressed in tobacco and potato plants and was shown to assemble into VLPs that are similar to the yeast derived commercial vaccine (see Fig. 1) (MASON et al. 1992; THANAVALA et al. 1995). The tobacco produced material was partially purified and used for parenteral immunization of mice. The resulting immune response was comparable to mice immunized with the commercial yeast-derived HBsAg. The limitation on this project has been the low level of expression of the particles in plant tissue, about 0.01% of the total soluble protein being recombinant HBsAg protein, approximately 1μg per gram fresh tuber using the tuber specific patatin promoter. With recent experiments it has been possible to increase the level of expression in potato tubers (L. Richter, manuscript in preparation). This will allow future feeding studies to test for oral immunogenicity.

3.3.2 Norwalk Virus

Norwalk virus infects humans and causes epidemics of acute gastrointestinal distress. There is no commercial vaccine although clinical trials are underway using VLPs composed of the Norwalk virus capsid protein (NVCP) produced and purified from baculovirus infected insect cell cultures (BALL et al. 1996). VLPs purified from insect cell cultures have been used for oral immunogenicity studies in mice. In a dose response study it was determined that mice require a minimum dose of 200μg NVCP for 100% of the mice to mount a significant immune response without using adjuvant. A significant immune response was defined as a serum antibody titer against NVCP at least fourfold higher than preimmunization titers.

Similar VLPs composed of NVCP have been produced in tobacco and potato (BALL et al. 1996). NVCP does not contain a signal sequence and Norwalk virions are produced in the cytoplasm of infected cells. Expression of recombinant NVCP in tissue culture tobacco leaves was up to 0.23% total soluble protein and 0.37% in potato tubers or 34μg per gram fresh tuber with the patatin promoter. Subcellular localization of the NVCP in the cytoplasm may allow higher expression levels than the previously described LT-B and HBsAg which are directed into the endoplasmic reticulum.

Oral immunogenicity studies have been performed by feeding tubers producing recombinant NVCP to mice (MASON et al. 1996). The mice developed both serum and mucosal antibodies comparable to mice immunized with VLPs from insect cell culture. In these studies, mice consumed four doses of 4g each of tuber containing 10–20μg NVCP per gram fresh tuber weight. Tuber slices were fed to mice either with or without 10μg of the adjuvant CT being added to the potato. Both groups of mice developed antibodies against NVCP. Although the mice fed samples with adjuvant had a higher titer of antibody response than those without adjuvant, this experiment shows it is not necessary to add adjuvant to the transgenic tubers expressing NVCP to induce an immune response. This is an important result since prior to this study, only LT-B tubers had been used for oral studies. LT-B is known to be a very immunogenic molecule and it was unknown whether other recombinant proteins would require addition of an adjuvant when transgenic tubers were fed to mice. NVCP provides an example of VLPs formed in plant material that can be fed directly to mice and results in an immune response without requiring adjuvant. When adjuvant was added, less recombinant NVCP was required to induce a significant immune response. Doses with only 10μg NVCP partially purified from tobacco and administered orally with 10μg CT were able to induce significant serum antibody responses. Studies such as this predict that coexpression of adjuvant and antigen within transgenic plant tissue may decrease the requirement for very high expression levels of the recombinant antigen.

Fecal IgA titers were assayed in the mice orally immunized with plant material containing NVCP (MASON et al. 1996). Although 32 out of 43 mice had significant serum titers against NVCP, only 12 out to 41 mice assayed were positive for fecal IgA specific for NVCP. A desired result of vaccines against diarrheal agents is induction of mucosal antibodies. Although serum titers were considered significant, they varied from 50 to 25,600 and for the animals responding in the lower range, the assay to measure sIgA may not have been sensitive enough to detect NVCP specific IgA. Alternatively, it may be necessary to use larger oral doses of NVCP to induce a measurable sIgA response.

3.3.3 Rabies Virus Glycoprotein in Tomato

Rabies virus glycoprotein (G-protein) has been expressed in baculovirus infected insect cells. Two doses of 100 and 200µg partially purified G-protein from the insect cells were given raccoons orally without addition of an adjuvant. Six out of seven immunized raccoons survived rabies virus challenge whereas all seven control nonimmunized raccoons died of rabies (FU et al. 1993).

As a first step in producing a plant based edible vaccine against rabies, McGarvey et al. (1995) created transgenic tomatoes expressing the rabies virus glycoprotein (G-protein). The recombinant G-protein was immunoprecipitated from both leaf and fruit tissue and was detected on western blots as two bands of 60 and 62.5kDa. The native G-protein from denatured rabies virus is 66kDa. The difference in sizes may be due to altered glycosylation and/or specific enzymatic cleavage by the plant tissue of sugar or amino acid residues. Results from immunoprecipitation and western blots demonstrate that important immunological epitopes are presented in the plant produced G-protein. The expression level of recombinant G-protein was estimated to be between 1 10ng/mg total soluble protein or 0.001%.

3.3.4 Other VLPs and Vaccine Antigens Expressed in Plants

There have been a few reports at meetings and in abstracts of scientific meetings of other VLPs and antigens produced in transgenic plants. A common problem has been the low level of expression of the foreign protein in plant material. As mentioned above, an oral dose of vaccine usually requires much more antigen than parenteral delivery for a given vaccine (CZERKINSKY and HOLMGREN 1995). It is necessary to empirically determine the required amount of antigen for a dose for each candidate vaccine. Delivery within plant tissue may change this amount to either a higher or lower dose than with purified antigens. These comparison experiments have yet to be performed. Advances in technology such as generating synthetic genes and information about expression in plant tissue will allow better expression levels for recombinant proteins. Also, future edible vaccines may be built into plants that already express an adjuvant such as LT (or a mutant, nontoxic form of LT as discussed in Sect. 3.1). This may allow much lower amounts of a second recombinant antigen for effective immune responses. When expressed for adjuvant effects, the level of recombinant LT or CT does not have to be very high, 10–25μg per dose (MASON et al. 1996; CLEMENTS 1996).

3.4 Autoantigens for Oral Immune Tolerance

Immune tolerance or hyporesponsiveness can be induced by oral presentation of soluble antigen (over 1 mg per dose) given continuously over time. Mechanisms of oral tolerance include cytokine-mediated active suppression, clonal anergy, and deletion of antigen-reactive peripheral T cells (CHEN et al. 1995; FRIEDMAN and WEINER 1994). Induction of immune tolerance is of great importance in autoimmune diseases such as diabetes but these type of studies require a constant supply of relatively large amounts of antigen. Using transgenic plants to produce the antigen in edible tissue, MA et al. (1997) have expressed glutamic acid decarboxylase (GAD), a diabetes-associated autoantigen, in low-alkyloid tobacco and potato. The recombinant GAD in tobacco leaf and potato tuber was estimated to be 0.4% of the total soluble protein for the high-level expressors. Plant material producing GAD (1–1.5mg GAD per dose) was added to the daily diet of young nonobese diabetic (NOD) mice starting from the age of five weeks until eight months old and the mice monitored for the onset of diabetes. Only 2 of 12 NOD mice fed the plant-produced GAD developed diabetes whereas eight out of twelve mice fed a control-plant diet developed diabetes. This is the first report of a recombinant autoantigen expressed in plant material being used to promote oral tolerance. This type of production of antigens in edible plant tissue would target a smaller audience than vaccines however, the constant high-level dosing requires a very large capacity of production. Edible transgenic plant tissue provides an ideal system for large scale production of autoantigens for daily food supplementation to induce oral tolerance.

4 Future Considerations

Through genetic engineering, vaccine antigens can be produced in edible plant tissue. By manipulating the expression constructs, the level of antigen can be increased to that which is feasible for human testing of oral vaccine doses. With further development of the technology, levels of recombinant antigen comprising 1% of total protein in edible tissue is a reasonable goal. For at least one commercially valuable protein, this level was exceeded, recombinant avidin protein is estimated at more than 2% total soluble extractable protein from dried transgenic maize seed (HOOD et al. 1997). Future considerations must include addressing potential problems in scaling up production, as well as distribution and handling of transgenic plant material. Identification of transgenic material, containment of the transgenes and control of recombinant protein may be potential problems for large scale production of vaccines in plants. Also, quality control for antigen content must be consistent. These considerations are particularly important since inappropriate dosing of vaccines can impair their effectiveness and constant dosing can potentially lead to immunological tolerance. Vaccines given to humans and animals need to be administered or overseen by the appropriate medical personnel.

Identification of transgenic plants may be accomplished by a visible marker. For banana it may be possible to engineer the pulp or peel of the banana fruit to be a red color instead of its normal color. In tomato, there are color mutations that might be used for identification that the fruit is not simply a "normal" tomato. Most commercial banana cultivars are sterile triploids that are propagated clonally. This eliminates the possibility of the transgene disseminating into wild species, however, bananas produce clonal daughter plants quite readily. One strategy may be to engineer the plant with an auxitrophic mutation which would require a nutrition supplement for growth at a certain stage of its life cycle to prevent undesired distribution of the plants. Some countries are using abandoned underground mines for growth of transgenic plants for containment and to prevent release of the "transgenes" into the environment.

Quality control of the antigen level produced in the edible plant tissues is important. Even though plants can be propagated clonally, the health of the plant throughout its lifetime can affect protein production and storage. Studies need to be performed to determine the consistency of clonal plants in production of recombinant protein. The level of recombinant avidin produced in transgenic maize varied twofold among field grown plants (HOOD et al. 1997). For oral vaccine doses there may be a threshold of antigen level that is a minimum for immune stimulation but a two to four fold increase would be within the effective dose range. This level must be determined for each vaccine. For example, mice require a minimum oral dose of 200µg NVCP for 100% of the mice to mount an immune response (BALL et al. 1996). A dose of 1 mg NVCP also promotes a desirable immune response therefore, as long as the minimum threshold of antigen level is obtained there is some allowable variation in production. This will need to be determined individually for every vaccine that is a candidate for production in plants.

Inducible promoters may allow better control over recombinant protein production levels in transgenic plants. Constitutive promoters may not be desirable with some antigens, for example LT-B affects the health of the plant at high expression levels. An ethylene inducible promoter would be useful for expression in banana or tomato. Many fruit ripening genes are ethylene inducible and several have been described in banana (CLENDENNEN and MAY 1997). With a promoter turned on at ripening, the fruit has already grown to its mature size so development of the edible tissue should not be adversely affected. Since ripening affects the color of many fruits, if there is a consistent correlation between color and antigen level it may be possible to use color as an indication of level of antigen to ensure an adequate dose of vaccine.

Vaccines must be taken in the appropriate regimen for that specific vaccine or the immune response may not be effective to immunize against the disease. The vaccine containing material should not be eaten in the quantity or frequency of food crops, hence there is a need for control over production and propagation. The concept of "edible vaccines" meets many of the criteria called for by The Children's Vaccine Initiative (MITCHELL et al. 1993); oral vaccines, subunit vaccines that can be combined, reduced cost of vaccines, and technology for developing countries to make their own vaccines. Developing countries have technology for vaccine production in transgenic plants: agriculture, harvesting and distribution. Production of edible vaccines may allow broader access to vaccinations by reducing the overall cost of the vaccination and also promoting compliance by the "user friendly" oral administration in receiving vaccinations.

Acknowledgements. We thank Charles J. Arntzen, Hugh S. Mason, Gregory D. May, Kenneth E. Palmer and Tsafrir S. Mor for editorial assistance.

References

Ball JM, Hardy MK, Conner ME, Opekun AA, Graham DY (1996) Recombinant Norwalk virus-like particles as an oral vaccine. Arch Virol Suppl 12:243–2491

Brandtzaeg P (1995) Basic mechanisms of mucosal immunity – a major adaptive defense system. Immunologist 3:89–96

Chen Y, Inobe J, Marks R, Gonnella P, Kuchroo VK, Weiner- HL (1995) Peripheral deletion of antigen-reactive T cells in oral tolerance. Nature 376:177–180

Clements JD, Sack DA, Harris JR, Van Loon F, Chakraborty J, Ahmed F, Rao MR, Khan MR, Yunus M, Huda N (1990) Field trial of oral cholera vaccines in Bangladesh: results from three-year follow-up. Lancet 335:270–273

Clendennen SK, May GD (1997) Differential gene expression in ripening banana fruit. Plant Physiol 115:463–469

Crawford SE, Labbe M, Cohen J, Burroughs MH, Zhou Y-J, Estes MK (1994) Characterization of virus-like particles produced by the expression of rotavirus capsid proteins in insect cells. J Virol 68:5945–5952

Curtiss R, Cardineau G (1989) Oral immunization by transgenic plants. Washington University, St. Louis, Patent Cooperation Treaty

Czerkinsky C, Holmgren J (1995) The mucosal immune system and prospects for anti-inflammatory vaccines. Immunologist 3:97–103

Danheiser SL (1997) Non-parenteral vaccine developers set sights on worldwide $4 billion market. *Genet Eng News* June, p 20

De Block M (1988) Genotype-independent leaf disc transformation of potato (*Solantim tuberostim*) using *Agrobacterium tumefaciens*. Theor Appl Genet 76:767–774

Dickinson BL, Clements JD (1996) Use of *Escherichia coli* heat labile enterotoxin as an oral adjuvant. In: Kiyono H, Ogra PL, McGhee JR (eds) Mucosal vaccines. Academic, San Diego, pp 73–101

Division of Microbiology and Infectious Diseases (1995) The Jordan report. NIH, Bethesda

Division of Microbiology and Infectious Diseases (1996) The Jordan report. NIH, Bethesda

Fernandez F, Conner M, Parwani A, Todhunter D, Smith K, Crawford S, Estes M, Saif L (1996) Isotype-specific antibody responses to rotavirus and virus proteins in cows inoculated with subunit vaccines composed of recombinant SA11 rotavirus core-like particles (CLP) or virus-like particles (VLP). Vaccine 14:1303–1312

Friedman A, Weiner HL (1994) Induction of anergy or active suppression following oral tolerance is determined by antigen dosage. Proc Natl Acad Sci USA 91:6688–6692

Fu ZF, Rupprecht CE, Dietzschold B, Saikumar P, Niu HS, Babka I, Wunner WH, Koprowski H (1993) Oral vaccination of racoons (*Procyon lotor*) with baculovirus-expressed rabies virus glycoprotein. Vaccine 11:925–928

Grangeot-Keros L, Enders G (1997) Evaluation of a new enzyme immunoassay based on recombinant rubella virus-like particles for detection of immunoglobulin M antibodies to rubella virus. J Clin Microbiol 35:398–401

Green K, Kapikian A, Valdesuso J, Sosnovtsev S, Treanor J, Lew J (1997) Expression and self-assembly of recombinant capsid protein from the antigenically distinct Hawaii human calicivirus. J Clin Microbiol 35:1909–1914

Greenspan N, Bona C (1993) Idiotypes: structure and immunogenicity. FASEB J 7:437 444

Haq TA, Mason HS, Clements JD, Arntzen CJ (1995) Oral immunization with a recombinant bacterial antigen produced in transgenic plants. Science 268:714–716

Hausdorff WP (1996) Prospects for the use of new vaccines in developing countries: cost is not the only impediment. Vaccine 14:1179–1186

Hood EE, Witcher DR, Maddock S, Meyer T, Baszczynski C, Bailey M, Flynn P, Register J, Marshall L, Bond D, Kullsek E, Kusnadi A, Evangilista R, Nikolov Z, Wooge C, Mehigh RJ, Hernan R, Kappel WK, Ritland D, Li CP, Howard JA (1997) Commercial production of avidin from transgenic maize: characterization of transformant, production, processing, extraction and purification. Mol Breeding 3:291–306

Huang S, Soldevila A, Webb B, Ghabrial S (1997) Expression, assembly, and proteolytic processing of Helminthosporium victoriae 190S totivirus capsid protein in insect cells. Virology 234:130 137

James C, Krattiger AF (1996) Global review of the field testing and commercialization of transgenic plants, 1986–1995: the first decade of crop biotechnology, vol 1. ISAAA Briefs. Ithaca

Kimbauer R, Chandrachud L, O'Neil B, Wagner E, Grindlay G, Armstrong A, McGarvie G, Schiller J, Lowy D, Campo M (1996) Virus-like particles of bovine papillomavirus type 4 in prophylactic and therapeutic immunization. Virology 219:37 44

Koletzki D, Zankl A, Gelderblom H, Meisel H, Dislers A, Borisova G, Pumpens P, Kruger D, Ulrich R (1997) Mosaic hepatitis B virus core particles allow insertion of extended foreign protein segments. J Gen Virol 78:2049 2053

Kuntz RM, Saltzman WM (1997) Polymeric controlled delivery for immunization. TIBTECH 15:364 369

Laurent S, Vautherot J, Madelaine M, Le Gall G, Rasschaert D (1994) Recombinant rabbit hemorrhagic disease virus capsid protein expressed in baculovirus self-assembles into viruslike particles and induces protection. J Virol 1994 68:6794–6798

Ma SW, Zhao DL, Yin ZQ, Mukherjee R, Singh B, Qin HY, Stiller CR, Jevnikar AM (1997) Transgenic plants expressing autoantigens fed to mice to induce oral immune tolerance. Nat Med 3:793 796

Mason H, Haq T, Clements J, Amtzen C (1998) Edible vaccine protects mice against E. coli heat-labile enterotoxin (LT): potatoes expressing a synthetic LT-B gene. Vaccine 16:1336-1343

Mason HS, Ball J, Shi J, Jiang X, Estes MK, Arntzen CJ (1996) Expression of Norwalk virus capsid protein in transgenic tobacco and potato and its oral immunogenicity in mice. Proc Natl Acad Sci USA 93:5335–5340

Mason HS, Haq TA, Richter L, Arntzen CJ (1996) Enhancing expression of foreign genes. Plant Physiol 111:40

Mason HS, Lam DMK, Arntzen CJ (1992) Expression of hepatitis B surface antigen in transgenic plants. Proc Natl Acad Sci USA 89:11745 11749

May GD, Afza R, Mason HS, Weicko A, Novak FJ, Arntzen CJ (1995) Generation of transgenic banana (Musa actiminata) plants via Agrobacterium-mediated transformation. BIO TECH 13:486 492

McCormick S (1991) Transformation of tomato with Agrobacterium tumefaciens. Plant Tissue Culture Manual B6:1 9

McGarvey PB, Hammond J, Dienelt MM, Hooper DC, Fu ZF, Dietzschold B, Koprowski H, Michaels FH (1995) Expression of the rabies virus glycoprotein in transgenic tomatoes. BIO TECH 13:1484 1487

Mitchell VS, Philipose NM, Sanford JP (eds) (1993) The Children's Vaccine Initiative achieving the vision. Institute of Medicine, National Academy Press, Washington

Nawrath C, Poirier Y, Sommerville C (1994) Targeting of the hydroxybutyrate biosynthetic pathway to the plastids of Arabidopsis thaliana results in high levels of polymer accumulation. Proc Natl Acad Sci USA 91:12760 12764

Newman T, Ohme-Takagi M, Taylor C, Green P (1993) DST sequences, highly conserved among plant SAUR genes, target reporter transcripts for rapid decay in tobacco. Plant Cell 5:701 714

Ohme-Takagi M, Taylor C, Newman T, Green P (1993) The effect of sequences with high AU content on mRNA stability in tobacco. Proc Natl Acad Sci USA 90:11811 11815

Perl A, Aviv D, Wilmitzer L, Galum E' (1991) *In vitro* tuberization of transgenic potatoes harboring beta-glucuronidase linked to a patatin promoter: effects of sucrose levels and photoperiod. Plant Sci 73:87–89

Perlak FJ, Fuchs RL, Dean DA, McPherson SL, Fischhoff DA (1991) Modification of the coding sequence enhances plant expression of insect control protein genes. Proc Natl Acad Sci USA 88:3324–3328

Rouwendal G, Mendes O, Wolbert E, de Boer A (1997) Enhanced expression in tobacco of the gene encoding green fluorescent protein by modification of its codon usage. Plant Mol Biol 33:989–999

Roy P, Bishop D, Howard S, Aitchison H, Erasmus B (1996) Recombinant baculovirus synthesized African horsesickness virus (AHSV) outer-capsid protein VP2 provides protection against virulent AHSV challenge. J Gen Virol 77:2053–2057

Roy P, Bishop D, LeBlois H, Erasmus B (1994) Long-lasting protection of sheep against bluetongue challenge after vaccination with virus-like particles: evidence for homologous and partial heterologous protection. Vaccine 12:805–811

Scerpella EG, Sanchez JL, Mathewson JJ, Torres-Cordero JV, Sadoff JC, Svennerholm AM, DuPont HL, Taylor DN, Ericsson CD (1995) Safety, immunogenicity, and protective efficacy of whole cell/recombinant B subunit (WC/rBS) oral cholera vaccine. J Trav Med 2:22–27

Sedlik C, Saron M, Sarraseca J, Casal I, Leclerc C (1997) Recombinant parvovirus-like particles as an antigen carrier: a novel nonreplicative exogenous antigen to elicit protective antiviral cytotoxic T cells. Proc Natl Acad Sci USA 94:7503–7508

Stover R, Simmonds N (1987) Fruit physiology, biochemistry and nutritional values. In: Wrigley G (ed) Bananas, 3rd edn. Longman Scientific and Technical, New York

Sturm A (1995) N-Glycosylation of plant proteins. In: Montreuil J, Vliefenthart, Schlacter H (eds) Glycoproteins. Elsevier, Amsterdam

Sugrue R, Fu J, Howe J, Chan Y (1997) Expression of the dengue virus structural proteins in *Pichia pastoris* leads to the generation of virus-like particles. J Gen Virol 78:1861–1866

Tacket CO, Mason HS, Losonsk G, Clements JD, Levine MM, Arntzen CJ (1998) Immunogenicity in humans of a recombinant bacterial antigen delivered in a transgenic potato. Nat Med 4:607–609

Taylor C, Green P (1995) Identification and characterization of genes with unstable transcripts (GUTS) in tobacco. Plant Mol Biol 28:27–38

Thanavala Y, Yang Y-F, Lyons P, Mason HS, ArntzenCJ (1995) Immunogenicity of transgenic plant-derived hepatitis B surface antigen. Proc Natl Acad Sci USA 92:3358–3361

Tsarev S, Tsareva T, Emerson S, Kapikian A, Ticehurst J, London W, Purcell R (1993) ELISA for antibody to hepatitis E virus (HEV) based on complete open-reading frame-2 protein expressed in insect cells: identification of HEV infection in primates. J Infect Dis 168:369–378

Wenzler H, Mignery G, Fisher L, Park W (1989) Analysis of a chimeric class-I patatin-GUS gene in transgenic potato plants: high-level expression in tubers and sucrose-inducible expression in cultured leaf and stem explants. Plant Mol Biol 12:41–50

World Health Organization (1995) World Health Report 1995. Geneva

Cowpea Mosaic Virus-Based Vaccines

G.P. Lomonossoff[1] and W.D.O. Hamilton[2]

1 Introduction

Short peptides (epitopes) have attracted considerable attention in recent years as potential sources of novel vaccines. Although the free peptides can often stimulate the production of antibodies, their immunogenicity can often be considerably enhanced by presenting them as multiple copies on the surface of a carrier protein. Because of this, a number of systems have been developed in which epitopes are genetically fused to a self-assembling macromolecule, often a virus particle (Lomonossoff and Johnson 1995, 1996). Among the particles which have been used for this purpose are those of plant viruses (Porta and Lomonossoff 1996; Scholtof et al. 1996; Johnson et al. 1997). The use of plant viruses for this application has a number of potential advantages: plant viruses often grow extremely well in their hosts and high yields of particles can be obtained; the purification of the particles is usually straightforward; many plant viruses are thermostable, raising the prospect of producing vaccines which might not require the refrigeration; the "reactor" used for the growth of the material i.e. the plant, is

[1] Department of Virus Research, John Innes Centre, Colney Lane, Norwich NR4 7UH, UK
[2] Axis Genetics plc, Babraham, Cambridge CB2 4AZ, UK

cheap and easy to maintain. In this chapter we describe the development of an epitope-presentation system based on cowpea mosaic virus (CPMV). By expressing heterologous sequences of surface of the virions, chimaeric virus particles (CVPs) are produced and we discuss the results from both immunogenicity and structural studies on such particles.

2 Development of a CPMV-Based Epitope-Presentation System

CPMV is the type member of the comovirus group of plant viruses. It infects a number of species of legumes and grows to particularly high titres in its natural host, cowpea (*Vigna unguiculata*). The genome of CPMV consists of two separately encapsidated positive-strand RNA molecules of 5889 (RNA 1) and 3481 (RNA 2) nucleotides. The RNAs each contain a single open reading frame and are expressed through the synthesis and subsequent processing of precursor polyproteins. While RNA 1 can replicate on its own in protoplasts, both RNAs are required to cause an infection in plants. RNA 1 encodes the proteins involved in the replication of both viral RNAs while RNA 2 encodes proteins involved in the cell-to-cell movement and encapsidation of the virus. Full-length cDNA clones of both genomic RNAs are available allowing manipulation of the CPMV genome.

The structure of CPMV particles has been solved to atomic resolution and a detailed description of the comovirus structure can be found in LOMONOSSOFF and JOHNSON (1991). CPMV capsids contain 60 copies each of a large (L) and a small (S) coat protein arranged with icosahedral symmetry. The two capsid proteins are folded into three anti-parallel β-barrel structures forming one icosahedral assymetric unit of the virus. In total, 180 β-barrel domains comprise the pseudo T = 3 (P = 3) structure of CPMV. Each L protein consists of two β-barrel domains which are positioned close to the icosahedral threefold axes. Five copies of the S protein, each consisting of a single β-barrel domain, are arranged around the fivefold axes. The strands of the β-barrels are linked to each other with loops of varying sequence; in the related animal picornaviruses, these loops commonly form the antigenic sites of the virus. An analysis of the three-dimensional structure of CPMV suggested that the loops between the β-strands would be suitable sites for the insertion of epitopes since these sequences are not involved in contacts between protein subunits. One of the loops of the S protein, the βB-βC loop, is highly exposed on the capsid surface and shows significant structural variation between different comoviruses (LOMONOSSOFF and JOHNSON 1995). Thus this site was initially chosen for the addition of epitopes. Although this chapter is concerned exclusively with the construction and properties of CPMV-based chimaeras containing insertions in the βB-βC loop of the S protein, recent work has shown that it is possible to insert heterologous sequences into other loops in the viral coat proteins while retaining virus viability.

The first CPMV chimaeras (USHA et al. 1993; PORTA et al. 1994) were constructed using clones containing full-length cDNA copies of the two viral RNAs downstream of a T7 promoter. RNA transcribed *in vitro* from these clones was infectious when inoculated on to cowpea plants. Subsequently, plasmids that contain full-length cDNA copies of RNAs 1 and 2 downstream of the cauliflower mosaic virus 35 S promoter were constructed (DESSENS and LOMONOSSOFF 1993). When linearised, these clones (termed pCP1 and pCP2) are directly infectious on cowpeas. These clones provide a cheaper, more efficient method for infecting plants than the use of *in vitro* transcripts and are now used routinely for the construction of chimaeras.

Early investigations into the use of CPMV particles to express epitopes determined certain guidelines for the construction of viable, genetically stable chimaeras (PORTA et al. 1994). Preferably, foreign sequences should be inserted as additions to the wild-type CPMV sequence and not used as replacements for native residues. Methods for introducing foreign sequences that result in sequence duplications either side of the insert may be unsuitable as loss of the insert can occur on passaging as a result of homologous recombination. The precise site of insertion can also have an important effect in maximising the growth of the chimaeras. Taking these guidelines into consideration, we have now refined a standard protocol for the introduction of foreign sequences into the βB-βC loop of the S protein (SPALL et al. 1997). This involves inserting DNA fragments encoding the foreign sequence of interest between the unique *Nhe*I and *Aat*II sites of a derivative of pCP2 engineered to contain an *Aat*II site in the βB-βC loop (DALSGAARD et al. 1997). Complementary oligonucleotides are synthesised which contain the sequence for the heterologous insert flanked by CPMV-specific residues and terminating in *Nhe*I and *Aat*II compatible ends. The oligonucleotides are ligated into the *Nhe*I/ *Aat*II-digested pCP2 derivative. This one-step cloning procedure allows the direct insertion of the foreign sequence without loss of any CPMV-specific residues. By this method the heterologous sequence can be placed at a number of different locations within the βB-βC loop of the S protein. For most chimaeras we have placed the inserts immediately upstream of Proline 23 of the S protein, an amino acid which is highly exposed on the surface of the CPMV capsid. To propagate the chimaeras, modified pCP2 (pCP2 chimaera) and pCP1 are linearised and mechanically inoculated on to cowpea plants. Once an infection is established (usually by 10 days post-inoculation), infected leaves are taken and the virus is extracted by the standard CPMV purification protocol. The method for the construction and propagation of CVPs is illustrated in Fig. 1.

Chimaeras containing inserts of up to 38 amino acids have been successfully made in which the presence of the heterologous sequence has not significantly affected the ability of the modified virus to replicate in plants. In most cases, the yields of CVPs are similar to those obtained with wild-type CPMV. The infections can be sap-transmitted to healthy plants allowing the efficient propagation of large quantities of CVPs. An investigation into the genetic stability of a number of chimaeras has determined that inserts can be maintained intact through at least 10 serial passages.

When purified CVPs are analysed by SDS-polyacrylamide gel electrophoresis, S proteins of the appropriately increased size are observed. In addition, a smaller protein (S'), which is not seen with wild-type virus, is frequently seen (Fig. 2). Amino-terminal sequence analysis of the S' protein from a number of chimaeras

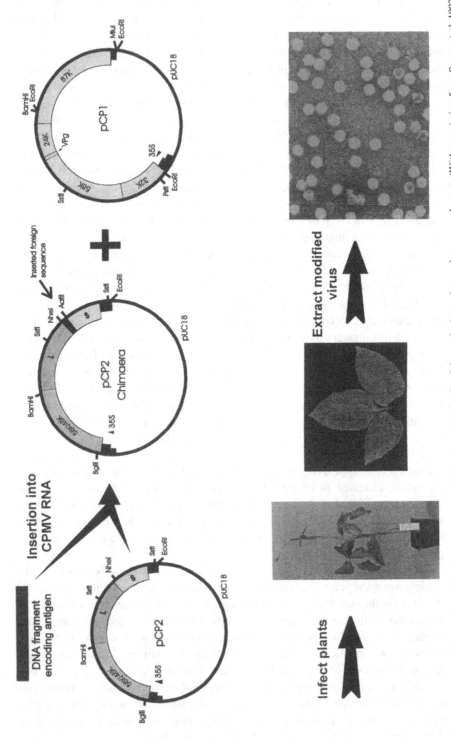

Fig. 1. Method of construction and propagation of CPMV chimaeras. For details of the methods used at each step, see the text. (With permission from SPALL et al. 1997)

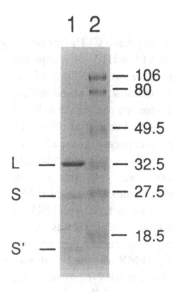

Fig. 2. Coomassie blue-stained SDS-polyacrylamide gel of the coat proteins of CPMV-HIV. *Lane 1*, SDS-denatured CPMV-HIV virions with the positions of the L and modified S proteins indicated on the left. S′ is the form of the S protein which results from cleavage near the C-terminus of the insert (see text). *Lane 2*, prestained protein markers (BioRad) with their molecular weights (*right*). (With permission from SPALL et al. 1997)

has shown that this protein arises due to a proteolytic cleavage event between the last two residues of the inserted sequence (MCLAIN et al. 1995; LIN et al. 1996). Structural studies (see Sect. 4) indicate that, at least in some cases, this cleavage does not result in the loss of the epitope from the surface of the virion but leads to it being anchored at only its N-terminus. This implies that a proportion of CVPs present the insert as a linear peptide rather than as a closed loop. This has potential implications for the immunological properties of CVPs.

3 Immunological Properties of CVPs

A considerable number of CPMV-based CVPs with inserts in the βB-βC loop of the S coat protein have now been successfully propagated and particles purified for immunological analysis. Candidate epitopes have been derived from human rhinovirus (HRV), human immunodeficiency virus (HIV), foot and mouth disease virus (FMDV), canine parvovirus, measles virus, *Staphylococcus aureus, Pseudomonas aeruginosa, Plasmodium falciparum* and other non-pathogenic agents such as hormones, tumor cell antigens, structural proteins and RGD-containing peptides. A requirement for any potential vaccine is obviously good immunological efficacy. The following section reviews the immunolgical data that has been obtained for several CVPs.

3.1 Human Rhinovirus 14

One of the first chimaeras to be examined immunologically was a CVP expressing a 14 amino acid sequence derived from residues 85–98 of VP1 of HRV-14, a sequence known as the NIm-1A site (SHERRY et al. 1986). When western blots of the chimaera, termed CPMV-HRV-II, were probed with an antiserum raised against HRV-14 virions, the denatured S protein carrying the epitope reacted strongly; no reaction was seen with S protein from wild-type CPMV (PORTA et al. 1994). To assess whether the particles were immunogenic as well as antigenic, a dose of 100μg the CVP in Freund's complete adjuvant (FCA) was used for intramuscular and subcutaneous immunisation of Dutch rabbits (PORTA et al. 1994). This was followed by several boosts using 50μg CVP in Freund's incomplete adjuvant (FIA). Good titres of antibody were obtained which were shown to react with HRV-14 VP1 on western blots using a serum dilution of 1:16,000. No reaction was seen to the HRV-14 VP2 or VP3 capsid proteins, nor was any cross-reaction seen with the HRV proteins when antisera raised against wild-type CPMV was used. Although the antibodies raised against the CVPs did react with intact HRV-14 particles, the binding was weak. These initial immunisations demonstrated that the epitope displayed on the surface of CPMV-HRV-II was capable of generating a strong specific immune response, at least against denatured HRV-14.

3.2 Human Immunodeficiency Virus Type 1

A considerable amount of immunological analysis has been carried out on a CVP, CPMV-HIV-III, containing a 22 amino acid consisting of residues 731–752 from gp41 of HIV type 1 (KENNEDY et al. 1986). Initially western blot analysis demonstrated that the denatured S protein of the CVP reacted with a monoclonal antibody specific for the HIV-1 insert (PORTA et al. 1994). Immunogold labelling studies with the same monoclonal antibody subsequently confirmed that the epitope was presented on the surface of intact CVPs (PORTA et al. 1996).

The first experiments to investigate the immunogenicity of the CPMV-HIV-III focused on the parenteral administration of purified CVP to C57/BL6 mice. The mice were injected subcutaneously with two 100 μg doses (35 days apart) of the CVP in alum (MCLAIN et al. 1995). Sera obtained 14 days after the boost gave a strong ELISA response (mean titre 1:25,800) when the peptide from gp41 was used in the assay. Antisera from control immunisations with wild-type CPMV did not react.

Neutralising titres to HIV-1 strain IIIB were determined using a quantitative syncytium-inhibition assay. At a 1:100 dilution, antisera from all of the mice were found to be neutralising with a mean value of 97% (MCLAIN et al. 1995). Antisera obtained at 14 days after a second boost at the 100 μg dose were shown not only to neutralise HIV-1 strain IIIB (92% at 1:100 dilution) but also neutralised strains RF (78%) and SF2 (66%). Further studies showed that mice strains C3H/He-mg and BALB/c also gave good ELISA and neutralising antibody responses; lowering the dose to 10μg CVP resulted in 99% neutralisation of HIV-1 at a 1:200 dilution (MCLAIN et al. 1996a). Doses of 1μg resulted in sera which neutralised virus to 97%, 79% and 63% from C3H/He-mg (H-2^k), C57/BL6 (H-2^b) and BALB/c (H-2^d) mice, respectively, indicating that neutralising antibody production elicited by the CVP did not have a narrow genetic restriction (MCLAIN et al. 1996b). Curiously, antisera raised against wild-type CPMV were also found to possess HIV-1 neutralising activity, although at a lower level that raised against the CVP, despite the fact that no anti-gp41 peptide antibodies could be detected. The HIV-1 neutralising antibody made against CPMV-HIV-III was shown to be HIV-1-specific by adsorbing the antisera to CPMV prior to carrying out the neutralisation assays (MCLAIN et al. 1995). The origin of the nonspecific neutralisation elicited by wild-type CPMV is unknown.

The longevity of the immune response in the C57/BL6 mice at 100 μg dose was relatively short-lived as the ELISA and neutralising antibody titres were no longer present at 48 days after the boost. However, 14 days after administering a second boost (i.e. a third injection), both ELISA and neutralising titres were found to be re-stimulated to levels similar to those after the first. The titres were again significantly reduced at 56 days after the second boost (MCLAIN et al. 1995). Another study showed that the longevity of the neutralising antibody response increased as the immunogen dose was decreased and appeared to be mouse strain-dependent (MCLAIN et al. 1996a). Using a dose of 1μg, a second boost improved the longevity of the neutralising antibody response in all three strains resulting in neutralisations of 95%, 98% and 73% 14 days after the third injection and 35%, 16% and 28% at 56 days, with C57/BL6, C3H/He-mg and BALB/c mice respectively (MCLAIN et al. 1996a). Comparison of the ELISA and neutralising titres after the first boost in C3H/He-mg mice showed that decreasing the dose from 100μg to 1μg resulted in a greater than 230-fold decrease in the ELISA titre whereas the neutralising titre only dropped 2-fold (MCLAIN et al. 1996a). Further studies using CPMV-HIV doses of 0.1μg and 0.01μg did not result in production of much neutralising antibody, even when three injections were used (MCLAIN et al. 1996b). The differential response of the ELISA and neutralising titres to a decrease in dose of CVP could indicate that the mice make HIV-1-specific antibodies to two distinct epitopes which are now known to be present on the gp41 peptide, one neutralising and one non-neutralising.

Studies have now been initiated which are aimed at evaluating the immuno-logical responses when CPMV-HIV-III is administered mucosally, either by the nasal or oral routes. Preliminary results have indicated that good antibody responses can be obtained after nasal administration with cholera toxin. Both IgG1 and IgA antibodies specific for the HIV gp41 peptide have been found in the sera and faecal extracts, respectively (T. McInerny, personal communication).

3.3 Canine Parvovirus

From the work with CVPs expressing the HIV gp41 epitope it is clear that the CPMV-based CVPs can be used to produce a strong immune response. The key demonstration of the utility of a CPMV-based vaccine, the ability to stimulate protective immunity, was recently reported in which a single administration of a CVP was shown to protect an animal from challenge by an infectious disease (DALSGAARD et al. 1997). In this study a CPMV-based CVP (CPMV-PARVO1) was constructed which contained a 17 amino acid epitope derived from the N-terminal region of the VP2 capsid protein of canine parvovirus. This peptide is also found in the capsids of mink enteritis virus and feline panleukopenia virus, all of which are part of the autonomous group of parvoviruses. CPMV-PARVO1 was administered subcutaneously as a single dose of 100μg or 1 mg into groups of mink, using an alum/Quil A adjuvant. Four weeks post-vaccination the mink were challenged oronasally with mink enteritis virus. Progression and severity of the clinical disease together with virus shedding was monitored. Protection was clearly afforded by either dose of the experimental vaccine and shedding of the mink enteritis virus was almost completely eliminated with the 1mg dose. Non-immunised mink became severely ill, whereas the CPMV-PARVO1 vaccinated mink were clinically healthy, maintained a good appetite and, with the exception of a single animal in the low-dose group, showed no signs of diarrhoea. This represents the first demonstration of an experimental vaccine produced exclusively by plants which confers protection against an infectious disease in the target animal. The actual amount of peptide given to the animals was calculated to be 3.5–7 times lower than the amount used in previous vaccination studies with mixtures of two overlapping synthetic peptides (LANGEVELD et al. 1995).

3.4 Other Chimaeras

Preliminary results of two other challenge studies have also been obtained. The first involves a pilot study using a CPMV-based CVP, CPMV-MV1, containing a 20 amino acid peptide corresponding to a T-cell epitope linked to a B-cell epitope (TB peptide), both derived from measles virus (OBEID et al. 1995). Using BALB/c mice, antibodies were raised against CPMV-MV1, which were shown to cross-react with measles virus. This antiserum was used in a passive transfer experiment whereby the serum was administered to naive mice prior to intracranial challenge with a neuro-adapted strain of measles virus. The mice were assessed 25 days post-challenge and a significant increase in the survival rate was found with mice receiving the anti-CPMV-MV1 antisera (57%) compared to the control group (12%; M. Steward personal communication). In contrast to these results, a similar study using the TB peptide gave no significant results, although a peptide in which the T-cell epitope was duplicated, the TTB peptide, did prove efficacious (OBEID et al. 1995). Although preliminary in nature, these results suggest that presentation of the TB peptide on the surface of CPMV may well provide enhanced immunogenicity.

The second pilot study has involved the construction of a CVP containing a 19 amino acid epitope from the G-H loop of FMDV VP1. Although made in significant quantities, this CVP has proved difficult to purify from plants as it shows a propensity to aggregate and form inclusion bodies in the plant cell. A partially purified preparation was made and each of eight guinea pigs inoculated with one or two 50μg doses of the CVP administered in FCA. Following challenge with FMDV, six animals showed a delayed onset of symptoms or partial protection and one animal appeared to be fully protected (A. King, personal communication). The protective effect of the CVP correlated with the amount of antibody induced.

4 Structural Analysis of Chimaeras

As the availability and conformation of a B-cell epitope is very important in eliciting a therapeutically relevant immune response, knowledge of the precise configuration of an expressed epitope would be of great benefit for engineering potential vaccines. To examine the conformation of an epitope presented on the surface of CPMV, the structure of the CPMV-HRV-II, the chimaera expressing a 14 amino acid sequence from VP1 of HRV-14 (described in Sect. 3.1) has been determined to 2.8Å resolution (LIN et al. 1996). This chimaera was chosen for the initial structural studies since the atomic resolution structure of HRV-14 is known (ROSSMANN et al. 1985) and it is therefore possible to compare the configurations of the epitope in its heterologous and native environments.

To solve the structure of the CPMV-HRV chimaera, crystals measuring 1–1.5 mm in diameter, were prepared under conditions similar to those used for the preparation of crystals of wild-type CPMV (PORTA et al. 1996). X-ray diffraction patterns were obtained with measurable reflections beyond 1.9Å allowing a high-resolution electron density map to be calculated. Since the crystals of CPMV-HRV-II had the same space group as those of wild-type CPMV, difference Fourier techniques could be used to solve the structure of the insert. Using this technique, all significant differences in density were found in the immediate vicinity of the βB-βC loop of the S protein indicating that the rest of the CPMV structure remained undisturbed by the insertion of the heterologous sequence. The electron density around the insertion site was weaker than that for residues not close to the βB-βC loop indicating that the loop is somewhat flexible. Nonetheless, the density was sufficiently above background to allow the polypeptide backbone to be traced precisely (Fig. 3). Thirteen of the HRV-14-specific residues were found to form a pseudo-closed loop on the surface of the chimaera. The cleavage event which occurs between the last two residues of the insert was evident in the structure of the chimaera, residues immediately following the cleavage site being clearly defined in the electron density map. The cleavage releases the penultimate residue of the insert allowing mobility and also enabling the loop to attain a relatively smooth conformation with a separation at the base of 19Å. In its native conformation, the NIm-1A epitope adopts a conformation including three sharp turns held by intra-loop hydrogen bonds. This conformation results from the constraint imposed by the short distance (9Å) between residues at the beginning and the end of the sequence. The conformational differences between the NIm-1A site as expressed on

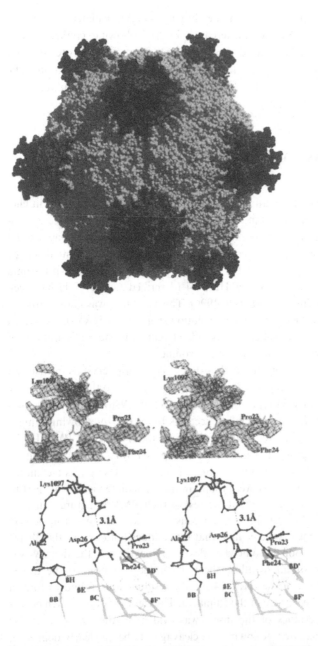

Fig. 3. Presentation of NIm-1A site of HRV-14 on the surface of CPMV. *Above*, space-filling model of CPMV-HRV particles. *Light grey*, L subunit; *dark grey*, S subunit; *black*, inserted epitope. *Below*, electron density for the chimaeric particle. The mainchain of the modified βB-βC loop was clearly defined in the averaged electron density map and sidechains of 12 of the 14 inserted residues were unambiguously modelled. The model is shown below the density; *light grey*, β strands and sidechains that comprise the immediate environment of Phe24. (With permission from LIN et al. 1996)

the surface of CPMV and in its native environment probably explain why antibodies raised against CPMV-HRV-II, while reacting well with denatured VP1 in western blots, bound only weakly to intact HRV-14 particles.

Additional chimaeras have been crystallised and are currently being examined crystallographically. By correlating the three-dimensional structures of the expressed peptides with their immunological properties it should be possible to optimise the way epitopes are presented on the surface of CPMV resulting in the production of chimaeras with optimal biological properties.

5 Future Developments and Commercial Prospects

The studies carried out so far clearly demonstrate the potential of CPMV-based CVPs as vaccines and provide a sound basis for the further development of a commercial vaccine based on this system, now referred to as EPICOAT. As with the development of all biopharmaceuticals, the amount of effort required to turn the results of a research programme into a commercial product should not be underestimated.

The development of pharmaceutical products by engineering plant viruses must meet all the usual regulatory and performance criteria for biopharmaceuticals before entry to human clinical trials. This requires the completion of pre-clinical studies undertaken in isolated tissues and in animals to generate data to support applications to the national regulatory authority for approvals to commence clinical trials. These studies include validation of the specific production and process technology, safety studies, proof of principle in animal models of the target disease and formulation and stability studies together with the implementation of appropriate quality control procedures.

The clinical trials process is comprised of three phases. Phase I clinical trials are normally conducted on small numbers of healthy human volunteers and are primarily designed to test product safety. Phase II trials are conducted to ascertain short-term safety and preliminary efficacy in a limited number of patients within the target population. Phase III trials are then conducted to evaluate comprehensively product safety and efficacy in patients with the target disease. Following satisfac-

Table 1. Typical time scales for the completion of the pharmaceutical product clinical approval procedure

Stage	Typical duration
Pre-clinical studies	6–18 months
Application for phase I trial	1–3 months
Phase I trial	6–12 months
Phase II trial	2–3 years
Phase III trial	2–3 years
Product licence approval	2–3 years
Product launch	1 years

tory completion of phase III trials a product licence application is submitted to national regulatory authorities requesting marketing authorisation for the pharmaceutical product. The entire process is strictly regulated and high standards are set by the regulatory authorities, the Food and Drug Administration (FDA) in the United States and the Medicines Control Agency (MCA) in the United Kingdom. Typical time scales for the completion of the pharmaceutical product clinical approval procedure are summarised in the Table 1.

Plants have been used extensively for the extraction of secondary metabolites, the products of which are easily qualified. Whereas suitable mammalian or bacterial cell culture methodologies have been developed for the production of recombinant pharmaceutical proteins, the use of plants as a production system for biopharmaceuticals has not. Plant leaves need to be established as a suitable raw material from which procedures can be developed to produce batches of purified CVPs in suitable quantities and of an appropriate quality, all under current Good Manufacturing Practice (cGMP) conditions, for subsequent testing in clinical trials. An appropriate formulation of the purified CVPs needs to be developed, in conjunction with a suitable adjuvant for parenteral administration, to facilitate the required immune response. Quality control procedures need to be developed and implemented in order to demonstrate the consistency and potency of the product. Specific issues regarding plant-derived pharmaceuticals have been recently reviewed by MIELE (1997).

The EPICOAT CVPs offer some distinct advantages as potential vaccines. The particulate nature of the CVPs makes them suitable for the generation of a mucosal, as well as systemic immune responses. The use of plants as a production vehicle offers advantages in terms of safety as the pathogenic agents of most concern in mammalian cell culture are not found in plants. As far as we are aware, there have been no reports of the pathogenic nature of plant viruses in man. Results from *in vitro* and *in vivo* studies commissioned by Axis Genetics have indicated that CPMV is not infectious to mammals and provided no evidence of pathological effects.

It is hoped that as a result of the efforts now underway an EPICOAT-based vaccine will enter the clinic within the near future and will subsequently result in the development of a marketable product having a clear competitive advantage for its target indication.

References

Dalsgaard K, Uttenthal Å, Jones TD, Xu F, Merryweather A, Hamilton WDO, Langeveld JPM, Boshuizen RS, Kamstrup S, Lomonossoff GP, Porta C, Vela C, Casal JI, Meloen RH, Rodgers PB (1997) Plant-derived vaccine protects target animals against a virus disease. Nat Biotechnol 15:248–252

Dessens JT, Lomonossoff GP (1993) Cauliflower mosaic virus 35S promoter-controlled DNA copies of cowpea mosaic virus are infectious on plants. J Gen Virol 74:889–892

Johnson JE, Lin T, Lomonossoff G (1997) Presentation of heterologous peptides on plant viruses: genetics, structure and function. Ann Rev Phytopathol 35:67–86

Kennedy RC, Henkel RD, Pauletti D, Allan JS, Lee TH, Essex M, Dreesman GR (1986) Antiserum to a synthetic peptide recognises the HTLV-III envelope glycoprotein. Science 231:1556–1559

Langeveld JPM, Kamstrup S, Uttenthal Å, Standbygaard B, Vela C, Dalsgaard K, Beekman N, Meloen R, Casal JI (1995) Full protection in minks against mink enteritis virus with new generation canine parvovirus vaccines based on synthetic peptide and recombinant protein. Vaccine 13:1033–1037

Lin T, Porta C, Lomonossoff G, Johnson JE (1996) Structure-based design of peptide presentation on a viral surface: the crystal structure of a plant/animal virus chimaera at 2.8Å resolution. Folding Design 1:179–187

Lomonossoff GP, Johnson JE (1991) The synthesis and structure of comovirus capsids. Prog Biophys Mol Biol 55:107–137

Lomonossoff GP, Johnson JE (1995) Eukaryotic viral expression systems for polypeptides. Semin Virol 6:257–267

Lomonossoff GP, Johnson JE (1996) Use of macromolecular assemblies as expression systems for peptides and synthetic vaccines. Curr Opin Struct Biol 6:176–182

McLain L, Porta C, Lomonossoff GP, Durrani Z, Dimmock NJ (1995) Human immunodeficiency virus type 1 neutralizing antibodies raised to a gp41 peptide expressed on the surface of a plant virus. AIDS Res Hum Retroviruses 11:327–334

McLain L, Durrani Z, Wisniewski LA, Porta C, Lomonossoff GP, Dimmock NJ (1996a) Stimulation of neutralising antibodies to human immunodeficiency virus type 1 in three strains of mice immunized with a 22 amino acid peptide peptide of gp41 expressed on the surface of a plant virus. Vaccine 14:799–810

McLain L, Durrani Z, Dimmock NJ, Wisniewski LA, Porta C, Lomonossoff GP (1996b) A plant virus-HIV-1 chimaera stimulates antibody that neutralizes HIV-1. In: Brown F, Burton DR, Collier J, Mekalanos J, Norrby E (eds) Vaccines 96. Cold Spring Harbor Laboratory, Cold Spring Harbor

Miele L (1997) Plants as bioreactors for biopharmaceuticals: regulatory considerations. Trends Biotechnol 15:45–50

Obeid OE, Partidos CB, Howard CR, Steward MW (1995) Protection against morbillivirus-induced encephalitis by immunization with a rationally designed synthetic peptide vaccine containing B- and T-cell epitopes from the fusion protein of measles virus. J Virol 69:1420–1428

Porta C, Lomonossoff GP (1996) Use of viral replicons for the expression of genes in plants. Mol Biotechnol 5:209–221

Porta C, Spall VE, Loveland J, Johnson JE, Barker PJ, Lomonossoff GP (1994) Development of cowpea mosaic virus as a high-yielding system for the presentation of foreign peptides. Virology 202:949–955

Porta C, Spall VE, Lin T, Johnson JE, Lomonossoff GP (1996) The development of cowpea mosaic virus as a potential source of novel vaccines. Intervirology 39:79–84

Rossmann MG, Arnold E, Erickson JW, Frankenberger EA, Griffith JP, Hecht HJ, Johnson JE, Kamer G, Luo M, Mosser AG, Rueckert RR, Sherry B, Vriend G (1985) Structure of a human common cold virus and functional relationship to other picornaviruses. Nature 317:145–153

Scholtof HB, Scholtof BG, Jackson AO (1996) Plant virus vectors for transient expression of foreign proteins in plants. Ann Rev Phytopathol 34:229–323

Sherry B, Mosser AG, Colonno RJ, Rueckert RR (1986) Use of monoclonal antibodies to identify four neutralisation immunogens on a common cold picornavirus, human rhinovirus 14. J Virol 57:246–257

Spall VE, Porta C, Taylor KM, Lin T, Johnson JE, Lomonossoff GP (1997) In: Shewry PR, Napier JA, Davis P (eds.) Antigen expression on the surface of a plant virus for vaccine production. Engineering crops for industrial end uses. Portland, London

Usha R, Rohll JB, Spall VE, Shanks M, Maule AJ, Johnson JE, Lomonossoff GP (1993) Expression of an animal virus antigenic site on the surface of a plant virus particle. Virology 197:366–374

Subject Index